W9-CWW-001

THE WORLD OF THE VIRUSES

By the Same Author

The Sea Inside Us
Civil War Medicine
Our Murdered Presidents
McBurney's Point

THE WORLD OF THE VIRUSES

by Stewart M. Brooks

South Brunswick and New York: A. S. Barnes and Company
London: Thomas Yoseloff Ltd

Library of Congress Catalogue Card Number: 70-88252

A.S. Barnes and Company, Inc.
Cranbury, New Jersey 08512

Thomas Yoseloff Ltd
108 New Bond Street
London W1Y OQX, England

First Printing March, 1970
Second Printing November, 1970

ISBN: 0-498-06983-4
Printed in the United States of America

Just as Alexander Fleming in 1928 would take note of a green mold destroying a culture of staphylococcus, Twort noticed that *clear* areas were developing in an otherwise cloudy plate of agar* containing billions of cells of the very same organism. Suspecting that this was a disease of bacteria, he transferred a little bit of agar from these clear areas (or "plaques," as they are now called) to fresh "staph" cultures and thereby induced new plaques. He then proved the infectious agent in the agar to be filterable and even showed that it multiplied and reproduced true to type. Twort's discovery had a tremendous impact on the thinking of the day, and the idea of germ against germ—with man the victor—served as the theme for Sinclair Lewis' novel *Arrowsmith.* And although this hope and dream was never to be realized in a practical way, the discovery caused another great stir in biology and proved a prime force in the development of modern virology.

In 1917 Felix d'Herelle of France confirmed Twort's observations—thereafter commonly known as the Twort-d'Herelle phenomenon—and coined the term bacteriophage (bacteria eater). And then d'Herelle went on to define a virus, to wit, "Viruses are particulate, invisible, self-reproducing, filterable entities that are obligate intracellular parasites." Which is to say, viruses are particles that can carry on only at the expense of some living organism and are always found *in* the cells (intracellular) of such an unfortunate organism. By contrast, in a bacterial infection the germs are always found outside the cells, not inside the cells. Thus, a virus is a parasite in the highest sense of the word.

In regard to d'Herelle's comments on viruses being "invisible" and "filterable," these no longer apply, for the

* A nutritional medium commonly used to culture bacteria and molds.

electron microscope makes the invisible visible, and filterability, as already noted, is a relative term. But basically the definition hits the nail effectively on the head because we are still obliged to consider viruses as particles which, on the one hand, exhibit the qualities of life and, on the other hand—when there is *no* living matter present—behave like laboratory chemicals in a bottle. In essence, then, viruses straddle the fence between the living and the nonliving, and this is not, as we shall see, a figure of speech.

Another startling revelation was published in 1911 by the American pathologist Francis Peyton Rous. Suspecting that a certain cancerous condition in chickens—now called Rous sarcoma—was of viral origin, Rous mashed up pieces of the diseased tissue in water and passed the mixture through ultrafine filters to screen out bacteria and other microscopically visible germs. And just as Iwanowski's filtrate had set off mosaic disease in healthy tobacco plants, so Rous's filtrate caused cancer in healthy chickens. Other deadly cancerous conditions of viral origin are myxomatosis of rabbits, mammary cancer in mice, a variety of leukemias of mice and chickens, and polyomas. The polyoma virus is unique among the cancer viruses because of the variety of tumors it induces and the wide range of animals attacked, including mice, rats, hamsters, rabbits, and guinea pigs. In mice the virus incites cancer of the bone, salivary glands, stomach, kidneys, and lungs.

Although the viral etiology of human cancer remains to be conclusively demonstrated, the case against that microbe grows stronger by the day. Already on the books are two kinds of *benign* tumor (cancer is a malignant tumor) known to be caused by viruses. One is the com-

To

Florence and Elward

Contents

Preface

BETWEEN THE LIVING AND THE DEAD ARE GALAXIES OF ANI-mated particles called viruses; no one knows whence they came or where they are headed. Viruses cause us great harm, and there is indeed every indication that in one way or another they are involved in cancer. On the plus side viruses have revolutionized our understanding of life and in the long run may prove more of a friend than an enemy. The purpose of this little book is to provide the interested reader with an educated general account of these twilight beings, especially as they relate to health and disease.

STEWART M. BROOKS
Auburndale, Mass.

Acknowledgments

I WISH TO THANK NATALIE ANN BROOKS, MY WONDERFUL wife, for all she has done to make this book possible. Too, I wish to express my appreciation to Mr. Thomas Yoseloff of A. S. Barnes & Co. for his confidence in my work. For the striking photographs which adorn this book I am deeply grateful to the National Medical Audiovisual Center; Merck Sharp and Dohme; Park, Davis and Company; Paul Black (Harvard University); William Burrows (University of Chicago); Jack Colvard (University of Maryland); Thomas Eisner (Cornell University); Fred Eiserling (University of California); Allan F. Howatson (University of Toronto); Joseph L. Melnick (Baylor University); John B. Miale (University of Miami); Councilman Morgan (Columbia University); Wendell M. Stanley (University of California); and D. A. J. Tyrrell (National Institute of Medical Research, England).

THE WORLD OF THE VIRUSES

1

Tobacco and Filters

EARLY IN THE STUDY OF THE CAUSE OF INFECTION IT BECAME quite clear that there were microbes that could neither be seen with the microscope nor cultivated except in *living* cells. In contrast, bacteria could be seen and in most instances cultivated in such lifeless materials as gelatin, sugar solutions, meat broths, and the like. The big trick would be to corral one of these invisible creatures and prove beyond the shadow of a doubt that it could cause a specific disease. And this was done in the year 1892 by a Russian botanist by the name of Dmitri Iwanowski.

Iwanowski's attention was turned to a strange abnormality of the tobacco plant known as mosaic disease, so-called because the leaves become a pattern, or mosaic, of healthy and discolored areas. When the leaves of a sick plant were ground up in water and a drop or two of the mixture added to the leaves of a healthy plant, Iwanowski proved in no uncertain way that the disease was infectious and that a microbe was indeed involved. And then he hit upon the ingenious idea of passing this infectious mixture through unglazed porcelain, a material with pores so small they cannot be seen with the ordinary microscope—and the *ordinary* microscope was the only microscope then available. Thus, any agent or microorganism that passed through such a filter would be ultramicroscopic—that is, beyond the power of the microscope to render visible.

And something did pass through, because when the filtrate was applied to healthy leaves the disease appeared. The way in which this fantastic discovery struck Iwanowski is interesting historically in itself, for he supposed that the microscopically invisible "something" that passed through the porcelain was simply an unusually small bacterium. This was certainly a reasonable and logical assumption for, after all, bacteria were known to be of various kinds and sizes. But six years later, in 1898, the Dutch botanist Martinus Beijerinck demonstrated that this invisible and filterable agent was not a bacterium. In the first place, it could diffuse through an agar gel; and in the second place, it could be precipitated by alcohol—properties diametrically opposed to those of ordinary living organisms. Put another way, diffusion and precipitation are nothing but well-known properties of plain, everyday chemicals. And so Beijerinck concluded that the cause of tobacco mosaic disease was a *contagium vivum fluidum,* or a "contagious living fluid." Today we call it a virus.

The work of Iwanowski and Beijerinck set off an investigational chain reaction of sorts in the laboratories throughout the world and at the turn of the century major discoveries and breakthroughs appeared on the horizon. In 1897 two German bacteriologists, Löffler and Frosch, filtered the fluid from the blisters in the mouths of cattle with foot-and-mouth disease and used the filtrate to transmit the virus to healthy animals; in 1900 the American Army Commission under Walter Reed isolated the virus of yellow fever; and in 1915 the English bacteriologist Frederick Twort discovered a virus that destroyed bacteria.

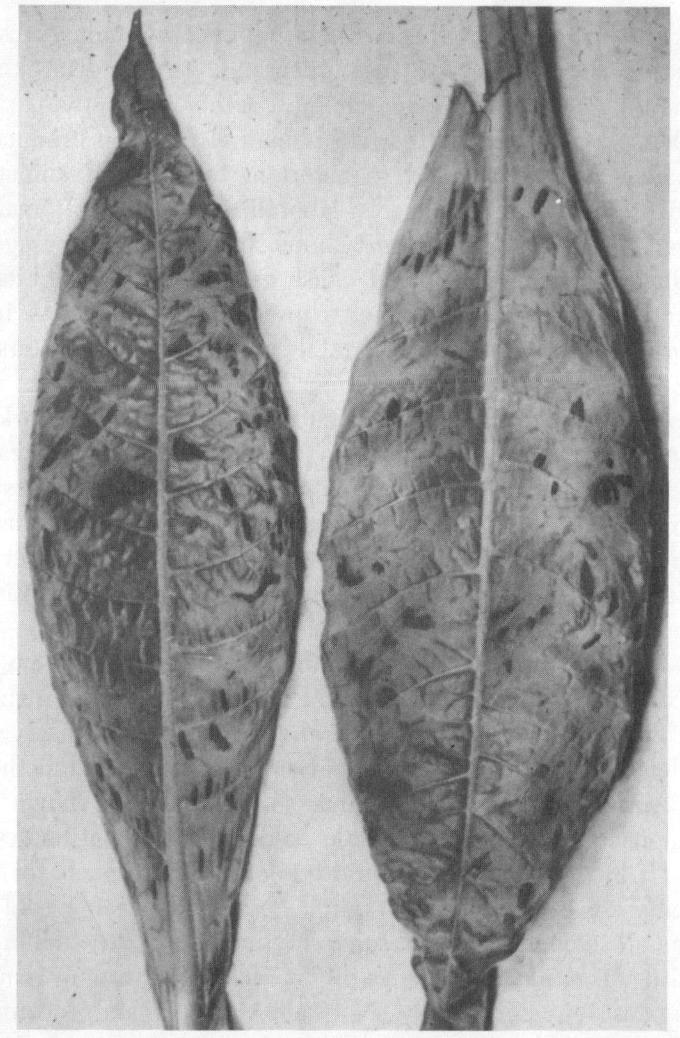

Tobacco mosaic disease. Darkish areas represent destroyed leaf tissue. (Courtesy CCM: General Biological, Inc., Chicago.)

mon wart and the other is molluscum contagiosum, a disease characterized by the appearance of roundish skin nodules, generally upon the face. Other suggestive evidence is the fact that certain human viruses can produce cancer in animals and that certain human and animal viruses can induce cancer in laboratory cultures of human cells. For instance, Richard Shope showed some time ago that the wart virus could cause cancer in wild rabbits, and recently a virus which causes a coldlike illness in man was shown to produce fatal cancer in baby hamsters. And perhaps the most exciting development relates to a cancerous condition in African children known as Burkitt's lymphoma. Not only do the involved cells reveal virus-like particles, but inoculation of monkeys with biopsy material has yielded evidence for the presence of a transmissible agent. Further, there is evidence that this very same virus may be the cause of infectious mononucleosis, a disease which we shall look at in more detail later on.

And more recently a number of investigators have called attention to some sort of relationship between the cold-sore virus and cancer of the cervix. No one has yet detected the virus in the crazily multiplying cells but the blood of practically all patients so far tested has shown a high concentration of telltale cold-sore virus antibodies.

If indeed cancer is of viral etiology the prospects for a cure are certainly encouraging. Possibly a vaccine could be developed or even some highly active virus-killing drug. Too, medical science might succeed in turning virus against virus. Recently, Norman Molomut and Morton Padnos of the Waldemar Medical Research Foundation discovered a virus that showed marked antimalignant properties by its ability to interfere with and prevent leukemia in mice which were purposefully injected with a leukemogenic virus; interestingly, this virus does not appear to produce any disease in several species of animals. If such findings are confirmed by other workers this affords new possibilities and hope.

2

Tadpoles and Spaceships

ONE OF THE MANY FASCINATING TWISTS IN THE HISTORY OF virology relates to the moment when the invisible became the visible. Or, put another way, just who was the first to see a virus? With few exceptions, textbooks either state or imply that a virus was not seen until the electron microscope came along in the late thirties. Actually, a Scotch physician by the name of Robert Buist saw virus particles —very large ones to be sure—five years before Iwanowski isolated the tobacco mosaic virus. What happened was this: In examining the cells taken from the vesicles, or skin eruptions, of patients with smallpox, Buist noted (under the microscope, of course) certain particles which he presumed to be related to the disease, particles which were not present in healthy cells. Other investigators soon reported the same phenomenon and before long these particles—soon to be referred to as "inclusion bodies"—became the hallmark of infection. To this very day their presence is considered of immediate significance in the diagnosis of smallpox, rabies, and certain other viral diseases.

Inclusion bodies in most instances are now believed to be aggregates of virus particles, a fact which was clearly

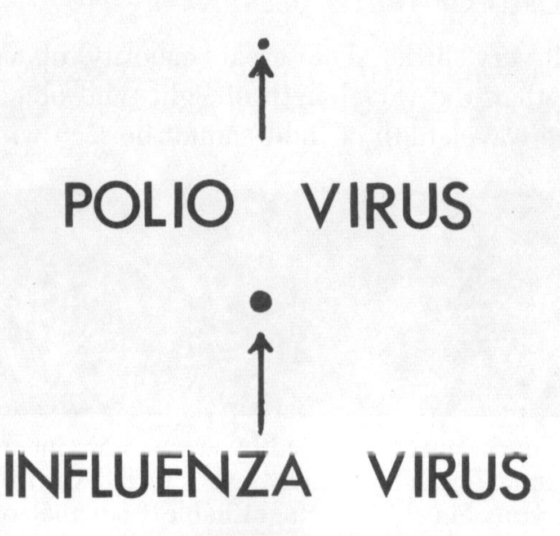

The size of two small viruses compared to a human red blood cell.

demonstrated over thirty years ago by the American pathologist Ernest Goodpasture. In a now classic experiment, Goodpasture teased apart the inclusion bodies of fowlpox (a viral disease of domestic fowl) into "ultramicroscopic granules," to use his words, which incited the disease in healthy birds. In short, Robert Buist was looking at aggregates of virus particles, or virions, as they are called today.

But Dr. Buist probably had little idea of what he was looking at, and even if he had he certainly would have learned very little. The great majority of viruses are smaller than the wavelength of light—and objects smaller than the wavelength of light cannot be seen with an ordi-

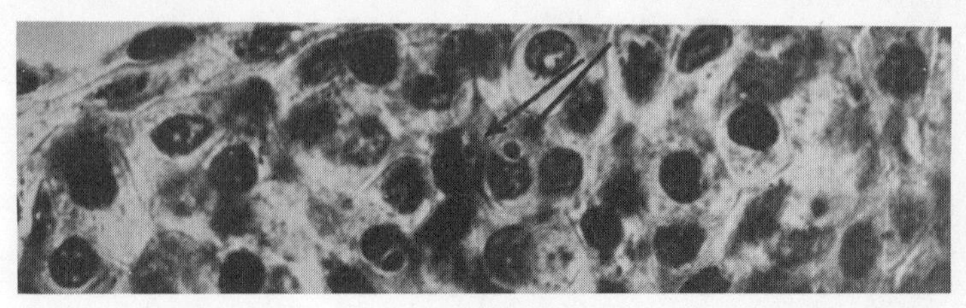

The two dark spots (indicated by arrows) are inclusion bodies in the corneal cells of a rabbit eye infected with the vaccinia virus, a virus almost indistinguishable from that of smallpox. (From Burrows, W.: *Textbook of Microbiology*, ed. 17, Philadelphia, 1959, W. B. Saunders Company.)

nary (light) microscope. By contrast, the electron microscope employs a beam of electrons whose wavelength is much smaller than the dimensions of a virus. Thus, not only does the instrument make viruses visible but it also has the power to magnify them several hundred thousand times. Actually, it is quite unbelievable. The foot-and-mouth virus, for instance, measures 23 millimicrons in size and so it would take over one million of them to span an inch (one inch = 25,000,000 millimicrons).

Viruses vary considerably in size, with the great majority falling between the wide limits of 10 and 200 millimicrons, a range, as indicated, far below the shortest wavelength of light. Too, the electron microscope discloses all sorts of shapes—rods, filaments, spheres, cubes, bricks, crystals, and even some that remind one of tadpoles and spaceships. Nonetheless, the basic structure of the virion is pretty much the same. In brief, it is composed of an outer protein coat, or capsid, wrapped around a central core of a highly complex chemical called nucleic acid. Typically, the capsid is divided into distinct individual subunits called capsomeres. Since a particular virion generally keeps its shape and has a set number of capsomeres, scientists can often identify a virus just by looking at it under the electron microscope. The tobacco mosaic virus, for instance, appears as an ear of corn with 2,130 kernels.

A great many viruses at first glance look spherical, but careful X-ray studies have shown that the capsid is actually an icosahedron; that is, a solid structure with 30 edges, 12 corners, and 20 faces, with each face divided into cap-

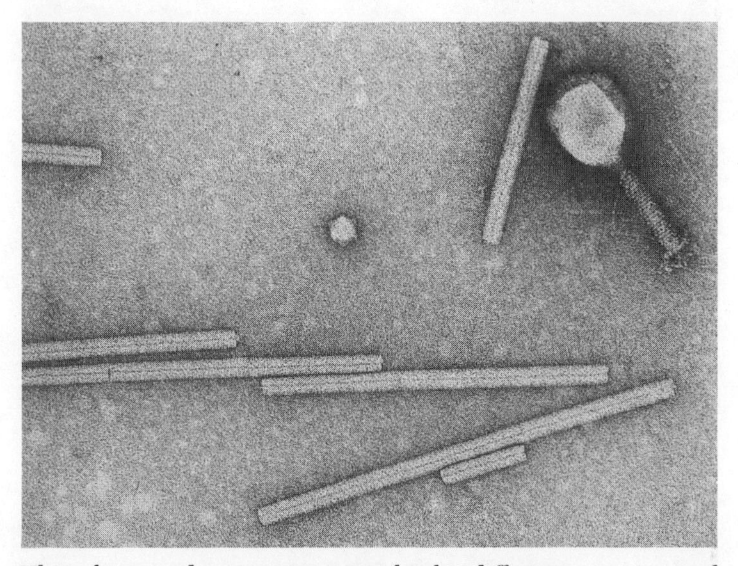

This photo underscores exquisitely the difference in size and shape among the viruses. Shown here—and viewed at a magnification of 280,000 times—are seven virions of the rodlike tobacco mosaic virus, and a virion each of the flask-shaped T4 and the spherical ϕX174 bacteriophages. (Courtesy F. A. Eiserling, Department of Bacteriology, University of California at Los Angeles.)

someres. The herpes simplex virus (the cause of the common cold sore), by way of example, shows an icosahedral arrangement with 12 capsomeres at the corners and 152 capsomeres distributed over the faces and along the edges. Viruses such as this—that is, those with edges, corners, and faces—are said to have cubic symmetry. Other viruses show either helical symmetry (capsomeres arranged in a helix or spiral form) or complex symmetry (a combination of cubic and helical). Finally, some virions are characteristically covered by some sort of gelatinous material referred to as the "limiting membrane" or envelope.

Perhaps the strangest and most interesting viruses are the so-called "T even" bacteriophages. In the language of virology they are virions of complex symmetry, but to most people they look very much like those Surveyor spacecraft we have been sending aloft to explore the surface of the moon. The head is shaped in the form of what geometry calls a prolate icosahedron (30 faces) and is attached at one end to a tail structure consisting of a contractile sheath surrounding a central hollow core. At the extreme end of the core there is a curious spiked plate carrying six slender tail fibers. The spikes and fibers affix the virus to a bacterial cell wall, whereupon the sheath contracts and drives the core through the wall—just like an injection.

And indeed, it is an injection, for once the needle or hollow core is in the bacterial cell the nucleic acid up in the head passes through the core and enters the cell. The events that follow are downright flabbergasting. First, the nucleic acid disappears and then in about ten minutes or so, lo and behold, a hundred or so virions appear from out of nowhere and cause the cell to rupture and burst forth its deadly contents. These virions attack other bacterial cells, and on and on it goes. Thus, a single virion can destroy billions of bacteria in a matter of hours. More or less, this is the way in which many viruses attack and destroy other living cells.

Recently, two American scientists, W. B. Wood and R. S. Edgar, were able to track down certain defective types of the T4 virus that produced half-baked offspring. That is, instead of the attacked bacterial cells spewing forth all normal, healthy virions, some were without heads, some were without tails, some were without tail fibers, and, in the most extreme situation, there was nothing but tail fibers. What Wood and Edgar did now was to separate the dismembered parts—heads, tails, and tail fibers—and combine them in a test tube. The result: They joined back together and produced normal, healthy virions!

3

Acids Alive

IN THE YEAR 1868, JOHANN MIESCHER, A SWISS PATHOLOGIST, performed an experiment that would eventually lead to the chemistry of life. And "eventually" is a rather tame adverb here because almost one hundred years elapsed before Miescher's findings were brought into sharp focus and acted upon. In a way, this reminds us of Dr. Buist in Scotland, the first man to see a virus.

Miescher became deeply interested in the chemical composition of the dark spherical body in the center of the cell called the nucleus. It was natural that the nucleus should have attracted the attention of biologists from the start, and it was also quite natural for them to assume that this very special body should play a key role in the workings of the cell. But chemistry was foreign territory to the biologist of a century ago, and it certainly is a stamp of Johann Miescher's advanced ideas that he should attack the nucleus in the test tube.

The pus that pours forth from an infected wound contains billions of scavenger cells called phagocytes. These cells are marked by the presence of large nuclei, and it was therefore to pus-rich discarded bandages that Miescher turned for his raw material. First, he concentrated the phagocytes and then he treated them with proteolytic enzymes, that is, enzymes which digest and dissolve away protein. Since all that remained in the wake of this drastic chemical treatment were cell nuclei, Miescher rightly concluded that the nucleus was mainly of a nonprotein nature. What is more, after further study he showed that this strange material represented a new class of biochemicals eventually to be called nucleic acids.

And here the subject of nucleic acids pretty much rested for a good fifty years because biologists and chemists alike were generally convinced that the key chemicals of the cell and life itself were proteins. Then, as history would have it, a seemingly slight error brought the subject of nucleic acids to front and center stage.

In 1935 Wendell M. Stanley, an American biochemist, isolated the virus of tobacco mosaic disease in the form of pure crystals, a feat that electrified the scientific world and won for Stanley the Nobel Prize in chemistry. The implications were startling—the "crystallization of life," it was said—but perhaps of equal impact on history was the fact that Stanley claimed the crystals were pure protein. Two years later, in 1937, Frederick Bawden and N. W. Pirie proved that the crystals were not pure protein. Instead, they were composed of 94 per cent protein and 6 *per cent nucleic acid!*

Now the pendulum of biological thinking started swinging the other way—toward the nucleic acids. What were these mysterious chemicals? What role did they play in the virus? Indeed, what role did they play in the cell? And so it happened that Stanley's error brought together the chemistry of the cell and the chemistry of the virus, a development that would eventually lead to unbelievable discoveries.

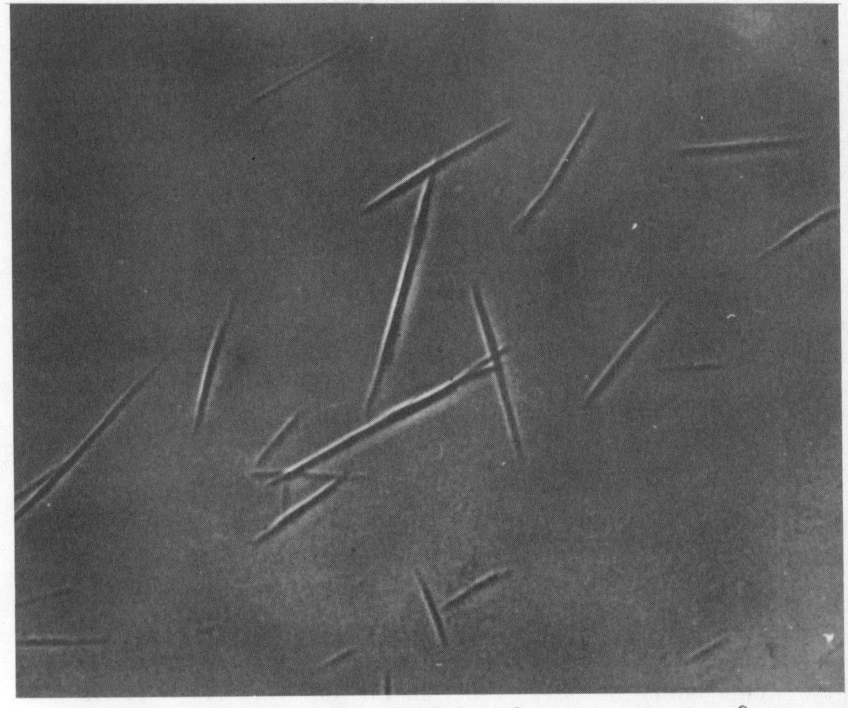

This historic photo of crystals of the tobacco mosaic virus first appeared in the *American Journal of Botany* (24:59, 1937). (Courtesy W. M. Stanley, Virus Laboratory, University of California at Berkeley.)

Nucleic acids are of two kinds. One is called deoxyribonucleic acid, or DNA, and the other, a very close relative, is called ribonucleic acid, or RNA. DNA is found in the nucleus of the cell and serves as a chemical brain in that it carries the blueprints for all cellular activities. Further, when the cell divides, the DNA in the two offspring cells is identical to that of the parent cell. That is to say, given the proper raw materials, a certain type of DNA is able to duplicate, or replicate, itself. And this has been demonstrated right in the test tube. In other words, DNA is "chemically alive."

Since a given type of DNA always carries the same set of chemical blueprints and since a cell owes its identity and character to its chemistry, we see that DNA is actually the hereditary material of the cell. Additionally, the gene —or the *unit* of heredity—may be thought of as a tiny segment of DNA.

Figuring out the way the myriad atoms of the DNA molecule are strung together ranks as one of the most spectacular achievements in chemistry and biology. From a clue turned up in 1950 by Erwin Chargaff and M. H. F. Wilkins, the English physicist Francis H. C. Crick and American biologist James Watson came up with the now famous Watson-Crick double helix. In essence, the molecule is made up of two very long chains of basic units (called nucleotides) wound helically around each other. More specifically, we may visualize the DNA molecule as two strings of colored beads (each bead representing a nucleotide) wound around each other in such a way that a bead in one string is delicately connected to a partner bead in the opposite string. And let us imagine that, when a cell divides, the two strings unwind, and that subsequently each in some way is able to bring about the stringing of loose beads (which happen to be lying about) into a new string which is identical to the old partner string. And this is possible, we must assume, because a red bead, for instance, will hook up only to a blue bead, and vice versa. Thus, the two offspring cells each end up with a DNA molecule just like the original. Consequently, the cells are alike.

In regard to the way DNA governs the cell, let us imagine that the two strings of beads uncoil, just as when the cell divides, and that each string lines up a partner string which this time turns out to be RNA. Once RNA is formed it then passes from the nucleus out into the cytoplasm, or cell proper, where it directs the synthesis, or building, of proteins. Some of these proteins are enzymes, that is, agents which determine the nature of all chemical reactions within the cell. Other proteins, along with materials such as fats and sugars, are forged into the fabric of the cell. In sum, an organism—be it a single cell, an alligator, or a man—is essentially an expression of the proteins which govern its chemistry and compose its cells and tissues. In turn, proteins owe their existence and identity to RNA and the latter in turn owes its existence and identity to DNA—the master molecule.

Returning now to the virus, it turned out that the nucleic acid of the tobacco mosaic virus was RNA. Soon DNA turned up in a virus, and before long it was established that all viruses are composed of *either* RNA or DNA wrapped up, as already noted, in a protein coat.

As to where viruses came from, there are two schools of thought. Some biologists regard them as "precellular" entities that appeared in the early evolution of life. Others consider that viruses have developed from bacteria in a sort of backward evolution. According to this latter view, nucleic acid made its way out of the bacterial cell and started off on its own. But whichever way we look at it, the virus is a twilight conglomeration of atoms. On the one hand, it is alive because it can reproduce itself—that is, the nucleic acid unwinds, and so on—and on the other hand, it is somewhat inanimate because it can only reproduce itself in the cells of *living* things. In a sense, a virus is like a contractor with a set of blueprints in his pocket. Given the necessary help and materials, he can build one house after another; without such assistance he is helpless. As one scientist put it, a virus is perhaps best thought of as a "piece of chemical information."

The great bulk of what we have learned about virus chemistry and nucleic acids stems from research involving a very obliging group of viruses we have already talked about—the bacteriophages, or "phages" as they are generally called today. The tadpole or spacecraft T4 phage, in particular, tells its story under the electron microscope and in the test tube in fantastic detail. Too, it tells it in an often surprisingly simple way. For instance, by placing T4 in a strong salt solution (which shrinks it) for several minutes and then diluting this as quickly as possible, the head bursts open and releases a tremendously long filament of pure T4 DNA.

Earlier we learned that the T4 virus attaches itself to the bacterial cell and gives its victim a shot of DNA, shortly after which the cell ruptures releasing a hundred or so new virions. From what has been said we can now readily understand the general idea of what goes on. T4 DNA is peculiar to the T4 virus, meaning that a shot of the stuff amounts to an injection of a set of chemical blueprints—a set of blueprints that will push aside the bacterial blueprints and take over the cell's chemical machinery for its own purposes. Imagine, if you will, a boot walking into a shoe factory, kicking out the boss, and then forcing the workers to make nothing but boots—enough boots to burst open the factory and spill over onto the sidewalks and streets.

Sometimes a T4 virus invades a bacterial cell and divides right along with it just as if nothing had happened.

As a matter of fact, such cells have an advantage over other bacteria in that they are immune to future T4 attacks! However, there may be trouble ahead, for occasionally an afflicted bacterial cell that can trace its ancestry back for a gigantic number of generations may suddenly get sick and in a matter of minutes burst forth with dozens of fresh T4 virions. This is especially likely to happen if such bacteria are exposed to ultraviolet light.

This phenomenon, called lysogeny, affords a remarkable insight into the understanding of heredity processes at the cellular level. Scientists believe that the T4 virus turns Dr. Jekyll inside the cell by actually combining with bacterial DNA, and in so doing is able to pass on and on for generations in perfect harmony. And then all of a sudden some influence comes along, such as ultraviolet light, to upset the chemical situation and turn Dr. Jekyll into Mr. Hyde.

But this is not all. A T4 virus may sometimes actually carry away from its mutilated host a tiny bit of bacterial DNA (hooked onto its own DNA), which means that when such a virus infects another bacterial cell it delivers two sets of blueprints. When this happens, and provided the virus plays Dr. Jekyll, the bacterial cell takes on certain characteristics of the former victim! Thus, a virus can act as a carrier of heredity for forms of life far removed from its own kind.

Another weird state of affairs arising from a partnership between a bacterium and virus concerns the disease diphtheria. Although the cause of this deadly malady is officially recognized as a bacterium with the fancy name of *Corynebacterium diphtheriae,* it has now been proved that a virus is also involved. The bacterium provokes the infection by releasing a powerful poison which acts upon

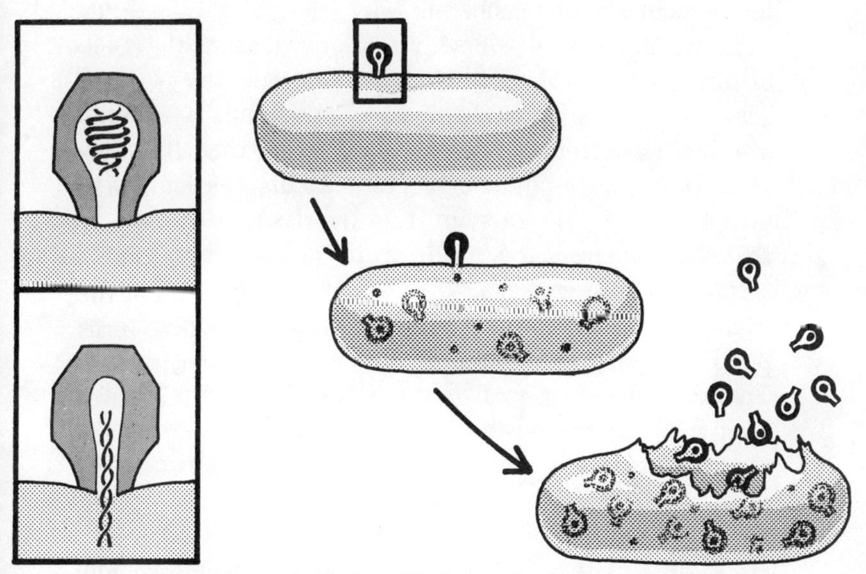

An artist's version of the T4 bacteriophage attacking and destroying a bacterial cell. (Courtesy of Eli Lilly and Company.)

the nervous system and vital organs. However, the bacterium produces the poison *only* after it has been invaded by a certain type of virus. Apparently, the bacterial DNA does not carry sufficient information to properly direct the cell's chemical machinery in this regard. Viral nucleic acid might very well supply the missing instructions.

When a virus enters an animal or plant cell the changes brought about depend upon the kind of cell invaded. Visible changes are generally of an injurious or destructive nature. Some host cells are literally "blown apart" (as underscored by phage action); some develop inclusion bodies (or colonies of virions); some become multinuclear (a normal cell has a single nucleus); and some start to multiply madly in a cancerous way.

These changes, of course, are for the most part signals of disease and explain what the doctor calls the patient's "clinical picture." For example, the signs and symptoms of rabies stem from the destructive changes that the rabies virus produces in the nerve cells. Again, the signs and symptoms of influenza stem from the destructive changes that the influenza virus produces in the cells that line the respiratory passageways. Viral cancer, too, is an excellent example, for the rapidly dividing cells develop into expanding tumors that generally overwhelm healthy tissues. There are, however, many viral diseases where the invader does not produce visible footprints, and here we must assume that the cellular alterations are of a functional rather than a structural nature.

An animal or plant virus may play Dr. Jekyll and Mr. Hyde, just as the T4 phage does. These so-called latent viruses are forever present, but seldom cause trouble because they are kept at bay by the forces of immunity. That is, they strike when our resistance is down, and a classic example is the cold-sore virus. And finally, there are viruses that apparently are harmless. Either that, or they are "looking for a disease."

If man could make a virus he would certainly be at the grass roots of molecular biology and life itself. On the practical side he would perhaps be on the road which one day might lead to the conquest of disease. For example, it is quite possible that a harmless synthetic virus could—via its DNA—transmit key cellular instructions relating to resistance and immunity. In short, if a bad virus carries bad information—and makes a cell sick—why not have a good virus carry good information and make a cell well?

After all, it is all a matter of DNA (and RNA) and essentially that is what this virus business is all about. No wonder, then, that in 1967 it was no less than the President of the United States who announced that Arthur Kornberg and his colleagues at the Stanford University School of Medicine had succeeded in the test-tube manufacture of viral DNA. Specifically, they synthesized the DNA of a bacterial virus named ϕX174 employing unbelievable chemical gymnastics. Just like the virus itself these workers caused an enzyme called polymerase to string together an array of nucleotides into a thriving double helix.

4

Eggs and Monkeys

MOST MICROBES GROW WELL ON SUCH LIFELESS FOODSTUFFS as meat broths, vegetables, gels, blood, and the like. "Blood agar," for example, commonly used for the culturing of many kinds of disease-producing bacteria, is simply and readily made by mixing together blood, agar, beef extract, peptone, salt, and water. This concoction contains a rich assortment of nutrients and, to boot, affords an excellent physical medium with which to work; that is, being a solid with the consistency of gelatin dessert (thanks to the agar) bacteria grow on the surface in discreet, easily seen colonies. All the bacteriologist has to do to get a sample for study is merely touch a colony with the point of a needle.

Unfortunately, viruses refuse to grow in or upon blood agar or any other lifeless medium. That they require living material—actively dividing cells—is a basic fact of microbiology in general and virology in particular. And historically this is a key point to remember, because one cannot study and learn about microbes without an ample supply, and many viruses were not really in ample supply until John F. Enders, Thomas H. Weller, and Frederick C. Robbins of Harvard University Medical School perfected the tissue-culture technique some twenty-five years ago. But this is getting ahead of our story.

In 1915, Frederick W. Twort of England, we shall recall, placed himself into the annals of history by noting that clear areas, or plaques, developed on an agar plate of staphylococcus, a phenomenon that countless bacteriologists had probably witnessed before but had not questioned. And apropos to our present interest, this discovery held the very essence of viral propagation—that is, a workable medium containing just the right diet for just the right virus, in this case a bacteriophage. Furthermore, since each plaque represents the destruction set off by a *single* virion, the method is of quantitative significance. For instance, if a drop of water causes the formation of twenty plaques when added to an agar plate of the appropriate species of bacteria, then we may conclude that the drop of water contained twenty virions.

But here, as indicated, the matter of growing viruses in dishes and flasks came more or less to a close for years to come. Bacteria would support phages, but obviously not other viruses. The only place to turn was to nature—to living plants and most especially animals. And actually what could be better? Plants and animals are made up of billions upon billions of cells and, what is more, they are nature's very own virus factories. Still, the obvious is sometimes the most difficult and this is certainly the case in the use of animals to culture viruses. The technical problems can be formidable and, in comparison, looking for a needle in a haystack may not be such a chore after all. Also, animals can prove tremendously expensive and sometimes hard to get. Again, animals are not equally susceptible and many different species are required.

But in the beginning, animals did serve as the virologist's test tube both for experimentation and for "bulk yields" in the making of vaccines. And today, too, animals are a *sine qua non* in research. Animals used include mice, hamsters, ferrets, rabbits, guinea pigs, and monkeys, among others. Suckling mice and wet chicks are especially useful because of their very low resistance to viral attack. For instance, whereas a twenty-day-old mouse is resistant to the virus of western equine encephalomyelitis, a six-day-old mouse readily succumbs to the disease. Even more impressive, a single virion will kill a four-to six hour-old wet chick.

The contribution of animals to virology notwithstanding, what was clearly needed was a setup whereby animal cells could be maintained alive and actively multiplying outside the body—that is, a tissue or, as some authorities prefer to call it, cell culture. This way, in essence, we would merely add the virus to the culture and watch it grow right before our eyes. Above all, the supply would be largely without limit.

The first breakthrough in this direction—indeed, the birth of the cell culture—was in 1907, the year that Ross G. Harrison of Yale University succeeded in keeping salamander cells continuously alive outside the body of the salamander. Six years later, Edna Steinhardt and her associates at Columbia University delivered the *coup de grâce* by propagating the cowpox virus in bits of cornea from the eye of a guinea pig. And then, somewhat strangely, in retrospect anyway, there was a pause of a good thirty-five years before anything of great significance came along to put the cell culture on the road to revolution.

But halfway through that pause, in 1931, a minor revolution occurred when Ernest W. Goodpasture and Alice M.

Woodruff of Vanderbilt University School of Medicine directed virology to the hen house. To wit, these workers were able to grow the virus of fowlpox in the embryonated, or fertile, egg. In a sense, nature had provided man with a sort of cell culture. The chick embryo and its associated membranes are constituted by a galaxy of rapidly multiplying cells, and this is just the kind of biochemical environment viruses prefer. Moreover, this environment is sterile—free of microorganisms—meaning that only the virus under study is involved. Finally, eggs are cheap and plentiful and, compared to animals, entail neither fuss nor bother.

Briefly, the procedure runs as follows: A fertile egg is incubated until about a week or so old and then by means of a syringe given a shot of the virus. This is done by making a window in the shell of the egg over the embryo or by drilling a hole in the shell. And it is most important to know exactly where to direct the needle, because whereas some viruses prefer the embryo proper, others do best in the yolk and still others in the allantoic sac. During this inoculating procedure, of course, the laboratory worker must exercise strict "aseptic technique" so as not to introduce any microbe into the egg other than the virus under study. This entails the full regalia of the operating room—gowns, face masks, sterile intruments, disinfectants, filtered air, and so on. Following inoculation the egg is incubated for two to three days, at which time the progeny of the inoculum is harvested. It has been estimated that during this time a single virion will generate an output of something like 100 million virions!

The fertile egg held sway up until the fifties, when John Enders and his colleagues perfected the tissue, or cell, culture technique for growing the virus of poliomy-

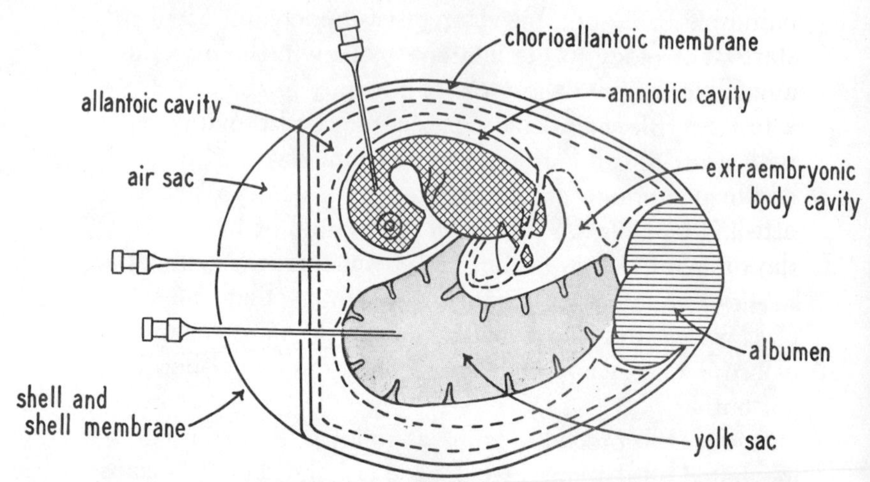

A schematic drawing of the embryonated hen's egg (10–12 days old) showing the chief modes of inoculating viruses for cultivation. (From Burrows, W.: *Textbook of Microbiology*, ed. 17, Philadelphia, W. B. Saunders Company, 1959.)

elitis. Frustratingly, the virus would not consistently multiply in mice, guinea pigs, or other small and abundant laboratory animals. And, as indicated, the egg would not work either. A paralytic infection in the monkey furnished the sole proof of the presence and identity of the virus. However, the monkey, by virtue of cost alone, was out of the question as a biological factory for the mass production of a vaccine. But let us not put the cost before the horse, because there is excellent reason for believing that a vaccine could never have been developed in the first place without Enders' tissue culture. One monkey now provides cells for hundreds of tests instead of just one, and monkey-tissue cultures can be shipped by air to any laboratory in the world.

Without letting the trees obstruct the forest, the tissue-culture technique is basically as follows: The culture is started by planting a few live tissue cells (human or animal—Enders originally used monkey kidney cells) in a test tube, bottle, or Petri dish and then covering them with a special nutrient fluid containing all the essentials of life plus an antibiotic for protection against bacterial attack (viruses are "immune" to antibiotics). In a few days the cells multiply to the extent of covering the floor

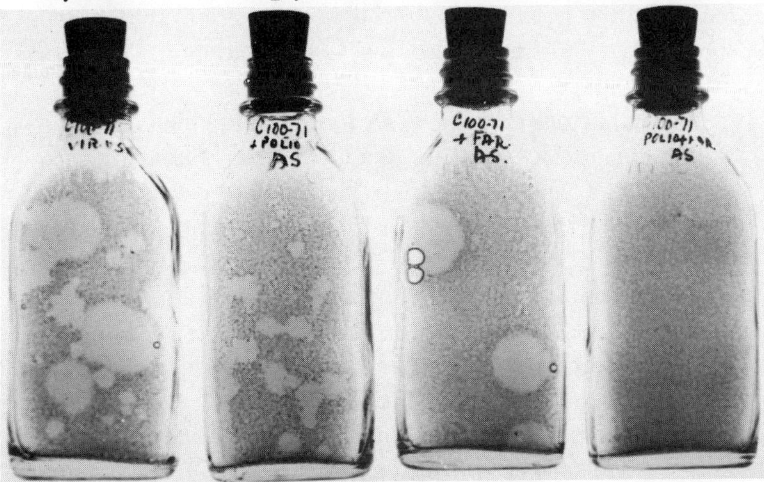

Bottle cultures of monkey kidney cells were used in detecting mixtures of viruses. The clear spaces (plaques) are areas in which the cells have been infected by virus and have degenerated. The bottle at the left contains two kinds of plaques: large ones characteristic of poliovirus, and small irregular ones typical of echovirus. In the second bottle, polio antiserum inhibited the growth of poliovirus, and only echovirus plaques appeared. In the third bottle, echovirus antiserum suppressed the growth of echovirus and permitted the isolation of poliovirus. In the bottle at the far right, antisera to both poliovirus and echovirus were present, and no plaques developed. (Courtesy M. Benyesh-Melnick and J. L. Melnick, Baylor University College of Medicine, Houston, Texas.)

or walls of the vessel with a "lawn" of tissue exactly one cell thick. Now the nutrient fluid is poured off and a few drops of virus suspension carefully spread over the "lawn." Finally, a thin layer of warm, melted agar is poured over the infected lawn to hold the virions in place. As the virus multiplies, the tissue cells are destroyed—the *cytopathic effect*—and visible and distinctive plaques appear. And herein lies the exquisite feature of the tissue culture; that is, not only do the plaques signal that a virus is present, but may also disclose its identity by their size and shape.

And what researchers have done to embellish and capitalize on the cytopathic effect is truly amazing. For example, let us consider the situation where a virus is *not* cytopathic. Here the virologist takes advantage of the fact that once a cell has been invaded by one virus it is typically immune to attack by another virus. So, if a sample suspected of containing a virus is first added to a tissue culture and then after an appropriate delay "challenged" by a virus of known cytopathic action, the absence of plaques constitutes a positive test. Contrariwise, if the sample did *not* contain a virus, the challenge virions would invade the cells and incite plaques (a negative test). Again, in the event a sample contains two or more viruses, *each* plaque may be taken to represent the progeny of a single virion, and to be *pure* for that particular virus. Thus, by transferring the material from single plaques to separate tissue cultures the original sample can be unraveled. Sometimes the researcher can tell how many kinds of viruses are present by taking note of the size and shape of the plaques. For example, if some are small neat holes in the lawn and the rest are large and "motheaten," the chances are good that two viruses are present. And again, blood serum containing antibodies against a cytopathic virus will prevent such a virus from developing plaques. For example, when the blood serum from a patient highly immune to all forms of polio is mixed with a polio virus, the latter is neutralized and fails to invade the cells of the tissue culture. The beauty of this, of course, is the factor of specificity; that is, polio antibodies neutralize polio viruses—and polio viruses *alone.*

And so modern virology was probably not born until 1949, the year that Enders, Weller, and Robbins reported their findings—and for which they received the Nobel Prize in Medicine. Writing in 1960, England's famed Christopher H. Andrews, the discoverer of the "cold virus," noted: "Largely as a result of Enders' work virologists in all parts of the world have in the last decade discovered more than three hundred new animal and human viruses."

5

Name, Please

THE CLASSIFICATION OF VIRUSES HAS PROVED A DIFFICULT task and to this day no one system satisfies everybody. The very simplest system, based on the kind of host, recognizes three huge categories: bacterial viruses, plant viruses, and animal viruses. Human viruses are considered animal viruses, but it is certainly interesting to note that some viral diseases occur exclusively in man and some exclusively in animals. Still, there are enough viral diseases common to both, such as rabies and encephalitis, to persuade one against laboring the point.

On a medical basis animal viruses are often classified in terms of the kind of tissue they normally attack in the body. According to this scheme there are five groups of viruses—dermotropic, pneumotropic, neurotropic, viscerotropic, and pantropic. Dermotropic viruses, as the term suggests, infect the skin (smallpox, chickenpox, measles, etc.); pneumotropic viruses infect the respiratory tract (common cold, influenza, etc.); neurotropic viruses infect the brain, spinal cord, and nerves (polio, encephalitis, rabies, etc.); viscerotropic viruses attack certain glands and abdominal organs (mumps, hepatitis, yellow fever, etc.); and pantropic viruses attack a variety of tissues.

The most elaborate and precise system of classification and the one preferred by most virologists takes into account a number of factors which, among others, include size and shape of the virion, type of nucleic acid (that is, DNA or RNA), and chemical behavior. On this basis eight major families of animal viruses are recognized. The salient points run as follows:

Poxviruses: These DNA viruses, the largest known, appear under the electron microscope as slightly rounded rectangular virions. Characteristically, they attack the skin—producing "pocks"—and the infected cells disclose inclusion bodies within the cytoplasm (that is, the cell exclusive of the nucleus). The so-called variolalike poxviruses cause smallpox (variola major), alastrim (variola minor), vaccinia, cowpox, mousepox, monkeypox, rabbitpox, pigpox, horsepox, goatpox, camelpox, canarypox, fowlpox, pigeonpox, and turkeypox. Additionally, a number of poxviruses incite tumors, such as rabbit fibroma, rabbit myxoma, squirrel fibroma, yaba monkey tumor, and molluscum contagiosum (in man).

Herpesviruses: These are large DNA virions (with cubic symmetry) surrounded by a limiting membrane or envelope. Studies carried out to date reveal the remarkable fact that all capsids of this family of viruses are fashioned into exactly 162 capsomeres. The chief members are the herpes simplex virus, the cause of the common cold sore, and the varicella-zoster virus (VZ virus), the cause of both chickenpox (varicella) and shingles (herpes zoster). The first invasion of the body by the VZ virus causes chickenpox, whereas shingles results from either the reinvasion of an immune host or the activation of a latent virus. A number of herpesviruses cause infections in animals. Monkeys display conditions similar to the common cold sore and shingles, but in horses and in chickens herpesviruses cause abortion and respiratory disease, respectively. Other animals subject to attack are the guinea pig, pig, cow, rabbit, dog, parrot, pigeon, and cormorant.

Interestingly, many herpesviruses undergo a marked change in virulence when they infect species other than their natural hosts. For instance, the very mild herpes simplex virus in mice and rabbits causes severe damage to the nervous system.

Adenoviruses: These are also DNA viruses, but they are considerably smaller, on the average, than poxviruses and herpesviruses. They are icosahedral in form (that is, resemble a polyhedron with twenty faces) and the capsid displays the phenomenal regularity of 252 capsomeres. The term adenovirus was coined in light of the historical fact that the first viruses of the family were isolated from the adenoids.

Of the 31 different kinds of adenoviruses, four (types 1, 2, 3, and 5) cause acute sore throat (with fever) in infants and young children and may remain in the tonsils and adenoids for long periods as latent agents of infection. Types 3, 4, 7, 14, and 21 cause respiratory illness in military recruits, with types 4 and 7 occurring in epidemic proportions in the United States. Type 8 produces an epidemic eye infection in adults and a disease in children marked by sore throat and conjunctivitis. In dogs, monkeys, cows, pigs, and birds, adenoviruses attack various organs. Recently, much excitement was created by the finding that the injection of human adenoviruses (types 7,

12, 18, and 31) into newborn hamsters caused the development of tumors.

Papovaviruses: This is the fourth and "final" family of DNA viruses. The papovaviruses are typically small compared to the other three families, but structurally resemble somewhat the herpesviruses and adenoviruses. With few exceptions these viruses are oncogenic (tumor producing) and incite benign and malignant growths in mice, rats, frogs, hamsters, squirrels, deer, rabbits, monkeys, goats, dogs, and horses. In man papovaviruses cause warts. Actually, the term papovavirus reflects the oncogenicity of the member viruses, to wit: *pa*pilloma *po*lyoma *va*cuolating virus. (Papilloma is a wart or similar skin growth; polyoma is a type of animal tumor; and vacuolating refers to the ability of many of these viruses to produce spaces—or vacuoles—within the infected cells.)

Myxoviruses: These are among the largest viruses containing RNA. The virions display helical symmetry (that is, the capsomeres are arranged in a helix) and most are surrounded by an envelope. Myxoviruses get their name from the fact that they have an affinity for mucin (a type of protein) on the surface of red blood cells. (The prefix "myxo" derives from the Greek word for mucus, the chief constituent of which is mucin.) The reaction is so pronounced that the red cells agglutinate or clump.

The myxoviruses are among the most vicious microbial enemies of man and animal. First and perhaps foremost is the influenza group that causes disease (influenza) in horses, swine, and birds as well as man. And another group, the paramyxoviruses, is represented by the mumps virus, parainfluenza viruses, and the Newcastle disease virus (one of the most economically important diseases of chickens). Other key myxoviruses are the measles virus, respiratory syncytial virus, and distemper virus. The respiratory syncytial virus, or RS virus as it is usually called, now turns out to be the most important single cause of pneumonia and bronchiolitis in infants and children. And the distemper virus, in addition to the dog, attacks the wolf, fox, coyote, weasel, ferret, mink, skunk, and raccoon.

Arboviruses: These are also RNA viruses, but much smaller than those belonging to the myxovirus family. They are of cubic symmetry and the capsomeres per virion number exactly 92. The cardinal and fascinating features of the arboviruses are that they infect the salivary glands and bodies of arthropods—namely, mosquitoes and ticks—without producing detectable tissue damage, and are transmitted to man or other hosts—mammals and birds—by means of an arthropod bite. This, incidentally, accounts for the prefix "arbo" (*ar*thropod *bo*rne).

There are about 125 arboviruses and apparently all are more or less pathogenic for man and animals. They are the cause of dengue, yellow fever, and almost endless forms of encephalitis (inflammation of the brain) and encephalomyelitis (inflammation of the brain and spinal cord). These include, by way of example, eastern equine encephalomyelitis, western equine encephalomyelitis, Venezuelan equine encephalomyelitis, Japanese B encephalitis, Murray Valley encephalitis, St. Louis encephalitis, and Russian spring-summer encephalitis.

Picornaviruses: The prefix "picorna" tells us two key facts about this family. First, they are among the smallest viruses (pico = "very small"), and secondly, they all contain RNA. Additionally, the virions are of cubic icosahedral symmetry with probably 32 capsomeres. The picornaviruses for the most part fall into one of two subgroups—the enteroviruses and rhinoviruses.

As indicated by the prefix ("entero-" means pertaining to the intestine), the enteroviruses are transient inhabitants of the gastrointestinal tract that may produce neural, respiratory, gastrointestinal, and eruptive skin diseases in man and such animals as pigs, cattle, rats, mice, cats, birds, and monkeys. The chief human members of the family include the polio viruses, Coxsackie viruses, and ECHO viruses. All three groups attack the nervous system and, to boot, Coxsackie and ECHO viruses may attack the heart and skin as well as the gastrointestinal tract. The Coxsackie viruses were named after the town in New York state where the first virus of the group was discovered during the early work on polio. As regards ECHO, virology had an acronymous holiday—to wit: *e*nteric *c*ytopathogenic *h*uman *o*rphan. Cytopathogenic (or cytopathic) refers to the fact that these human enteroviruses produce visible destruction of cells, and "orphan" was tagged on because at the time of their discovery they did not appear to be related to any disease. Today we know that ECHO viruses are anything but orphans, and those that are we may assume are trying hard to find a disease.

The rhinoviruses compose a relatively new group of picornaviruses recovered from human cases of common cold, croup, bronchitis, and bronchopneumonia, and have been shown to produce colds in volunteers. As it stands and until proved otherwise this group represents the ubiquitous "cold virus," a subject we shall go into in some detail a little later.

A picornavirus somewhat in a class by itself is the foot-and-mouth virus, the first animal virus to be discovered. Different types of this virus—the smallest known, incidentally—vary in their proclivity for a particular species. Cloven-footed animals are naturally susceptible to foot-and-mouth disease, with cattle, swine, sheep, and goats the most often affected. Horses have been known to get the disease, however, and a few cases in man have been authenticated.

Foot-and-mouth disease is also commonly called *hoof*-and-mouth disease, and perhaps the latter is actually more appropriate because there is now on the books a human involvement called "hand-*foot*-and-*mouth* disease." This is a vesicular eruption of the mouth, hands and feet caused by three types of Coxsackie viruses. The illness, which was apparently widespread in the summer and fall of 1966, is usually mild but severe infections, such as meningitis and a paralytic disease resembling polio, have been reported. Even death due to myocarditis (infection of the heart muscle) has been described.

Reoviruses: This is a rather new family of RNA viruses isolated from the feces and respiratory tract of man and a variety of domestic and wild animals, including monkeys, chimpanzees, cattle, dogs, and mice. But despite the presence of these viruses in children ill with diarrhea and respiratory diseases, virologists, at this time anyway, have not been able to establish an etiologic relationship. As it stands, therefore, these viruses remain quite true to their acronym "reo-" (*r*espiratory *e*nteric *o*rphan).

6
Attack

WE LIVE IN A MICROBIAL SEA, AND WHETHER OR NOT WE SINK or swim depends upon the body's resistance to infection. And nowhere is Nature more ingenious than in the myriad ways and means she employs to fight the microbe. A great deal about these ways and means is known, but at this point it would be highly naïve to assume that we have anywhere near the complete picture. Alas, much research serves essentially to embellish our ignorance. Why, for instance, is the dog susceptible to distemper and not to polio, and why is man susceptible to polio and not to distemper? Why, for instance, do we have mumps only once and cold sores over and over again? Why, for instance, can an attack of measles head off an attack of chickenpox and vice versa? Why, for instance, are run-down individuals less susceptible to viral attacks than the well-fed and robust? And so on.

When a virus enters the body it may or may not cause trouble, depending upon the nature of the virus and the state of resistance. Some viruses apparently are just along for the ride, while others will incite an infection if they are able to overcome those "ways and means" mentioned earlier. The precise mechanism of infection still remains somewhat of a mystery. Some viruses produce visible cellular damage whereas others do not. And even in the latter instance, one may well ask what is the "mechanism of the damage." For the present, about the best we can do is to say that a virus makes us sick by invading the cells of the body and taking over the chemical machinery for its own reproductive purposes.

And this brings up the matter of which cells—all the cells or just certain cells? All the cells might very well be involved—indeed, an infection commonly makes us "sick all over"—but usually some suffer more than others, a fact which explains the highly characteristic signs and symptoms of the common cold, measles, mumps, and chickenpox, to name a few. Actually, as we learned a while back, this affinity of a particular agent for a particular tissue or organ is the basis for classifying viruses as dermotropic, pneumotropic, neurotropic, and so on. This is all well and good provided we appreciate the fact that a virus may suddenly change character. The chickenpox virus, for instance, sometimes attacks the brain.

The first-line defenses against a microbial attack are the skin and mucous membranes, both of which, in man, serve as physical barriers. The skin is the most resistant of the two and with few exceptions can effectively keep out just about any microbe; the abraded skin, of course, is a hole in the dike. The mucous membranes, which line the gastrointestinal and respiratory passageways, are much more delicate than the skin and not uncommonly are penetrated by certain microorganisms even when intact. A special feature of the respiratory mucous membranes is the microscopic hairlike structures, called cilia, which by their continuous forward-beating movement "bring up" the mucus secreted by the mucosal cells. Thus, the mi-

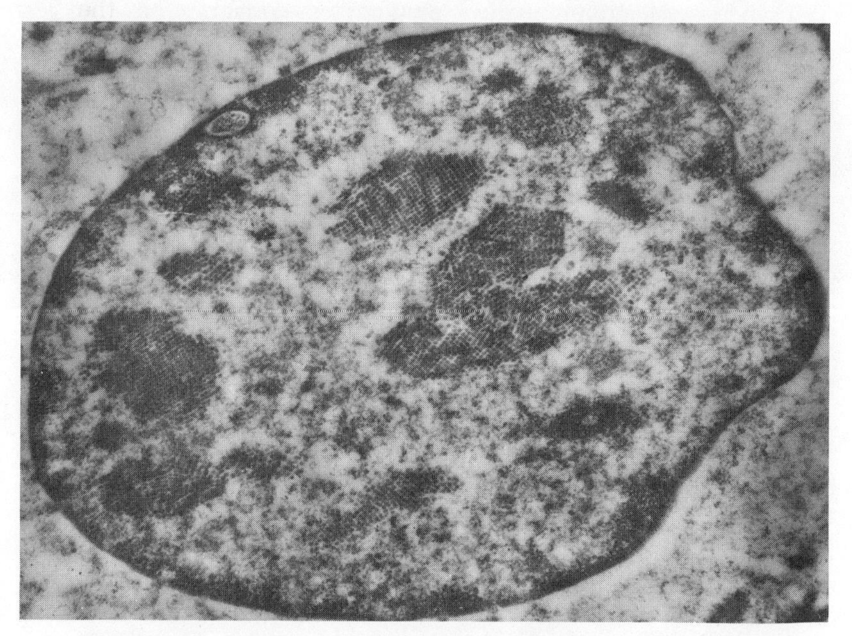

Nucleus of a cell from a wart (magnified 23,000 times) showing wart virions agglomerated into ten or more "crystals." Note especially the orderly arrangement of the virions in each crystal. (Courtesy Allan F. Howotson and the *Journal of Investigative Dermatology.*)

crobes present in the air we breathe become trapped in the sticky stuff and are eventually swallowed. This reminds us a little of robbing Peter to pay Paul, but the setup makes sense because the extremely high acidity of the gastric juice serves as a most effective disinfectant.

The second-line defenses come into play to take care of those microbes which invade the blood and other tissues. The white cells of the blood and related cells elsewhere—collectively called phagocytes—converge on the invaders and proceed literally to "eat them up." (This is just what the word phagocyte means—a "cell that eats.") The action (called phagocytosis) reaches its acme in areas of inflammation where the phagocytes accumulate as pus. And above all, especially where viruses are concerned, there is a galaxy of antimicrobial chemicals, of which *antibody* and *interferon* appear to be of chief concern.

Discovered in the mid 1950s by a British virologist, the late Alick Isaacs, interferon is a relatively small protein produced by a cell once it has been invaded by a virus. Interestingly, this agent neither neutralizes nor destroys but it does, as its name suggests, interfere with viral replication. This explains the perplexing observation that when a cell is infected with one virus it frequently cannot be infected with another virus. For more than a decade now researchers have been trying to harness this action for medical purposes. A number of avenues have been explored but the only promising lead—and a very good one—centers around the use of certain chemicals to stimulate the cells to produce interferon. Recently it was shown that a shot of synthetic ribonucleic acid (RNA) stimulated enough interferon to cure a serious eye infection in rabbits. Perhaps one day the chemist will come across a similar type chemical that will serve as a universal vaccine.

Antibody, from what we know at present, appears to be the body's most important defense against viral attack. Unlike interferon, it is highly *specific* in its action; that is, chickenpox antibody neutralizes the chickenpox virus and that virus alone. Thus, we speak of antibod*ies*, with a particular virus stimulating the production of a particular antibody. The exact mechanism by which this is done resides in the realm of pure theory. Suffice it here to say that a virus is an *antigen*—that is, an agent chemically foreign to the body—and like all antigens stimulates the plasma cells located in lymph nodes (and elsewhere) to produce and release gamma globulin proteins, all of which serve as antibodies. In a word, then, viruses are *antigenic*—and the majority are highly antigenic.

Once stimulated, plasma cells continue to release anti-

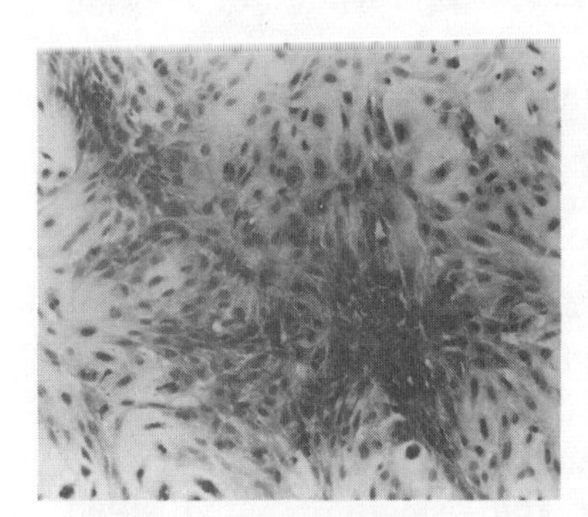

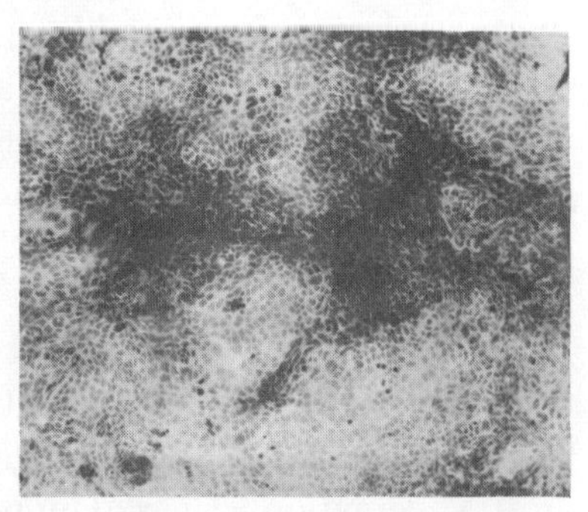

Normal mouse kidney cells (left) are markedly transformed (right) by the cancer-producing virus known as simian virus 40. (Courtesy Paul H. Black, M.D. and *The Journal of the American Medical Association.*)

body and, what is more, step up their output upon subsequent encounters with the virus. Though the initial stimulation does not produce sufficient antibody to quell the virus and prevent infection, subsequent encounters, as indicated, cause no harm. Thus, with but very rare exceptions we suffer measles only once even though we encounter the virus throughout life.

But what about viral diseases, such as the common cold and flu, that we have over and over again? Actually, these infections, technically at least, are not exceptions, for there are several types of cold virus and several types of influenza virus. Moreover, some viruses, the influenza virus, for instance, may change character (*mutate*) overnight, meaning that a relatively harmless virus could turn killer—a killer against which the body has not had an opportunity to build up antibody.

The newborn receives a heavy dose of antibodies from the mother's blood, and therefore is protected for the first two or three months of life. Also, if breast-feeding is used, the baby receives additional antibodies in the mother's milk. As the days, months, and years pass by the plasma cells are stimulated by this and that microbe and by adulthood the blood and other body fluids contain a galaxy of antibodies. We have measles and mumps and chickenpox, and then forevermore are rid of them. Ideally, of course, we would prefer to build up this antibody defense without suffering this or that disease, and to a large extent this is possible through vaccination.

Vaccination is based upon the principle that a killed or attenuated (weakened) microbe may stimulate the plasma cells to the same degree as the living, "wild" microbe. A "killed vaccine" is obviously preferred, but many organisms when killed (by heat or chemicals) are so drastically altered that they lose their ability to stimulate the plasma cells. This is particularly true of viruses, which means that effective viral vaccines in most instances must be prepared from the attenuated agent. In brief, such a vaccine is usually prepared by repeated culturings of the wild virus in tissue cultures, eggs, or even in animals. In some way, as yet unknown, the virus loses its virulence but not its ability to stimulate plasma cells. The procedure is much like transferring a mean tiger from cage to cage, allowing him to spend a few days in each; finally the tiger emerges tame as a kitten.

This tiger analogy is especially apropos because the tame kitty one day might possibly leap at somebody's throat, and virologists themselves cannot say with absolute certainty that an attenuated virus might not act up in the same way. In the case of polio, a discussion is still going on relative to the use of the killed (Salk) vaccine versus the attenuated or "live" vaccine (the one given orally). All tests show a much greater antibody response to the attenuated vaccine, but a few cases have been reported in which this vaccine has been implicated as the cause of the disease. We shall have more to say about this later.

Another approach in the making of a vaccine is to use a virus chemically and biologically similar to the one that causes the disease. The classic example here is the use of vaccinia virus—a virus that causes a very mild disease in man and cattle—to provoke immunity against the deadly smallpox. Apparently, the vaccinia virus provokes the plasma cells to produce the same type of antibody as the smallpox virus itself.

Vaccination, of course, involves time, that is, time for the plasma cells to produce sufficient antibody to thwart the invader, and therefore the procedure is usually of no avail in nonimmune persons who become suddenly exposed. In this eventuality, gamma globulin—the antibody "fraction" derived from the blood of immune individuals—affords short term aid in the prevention of polio, measles, mumps, and infectious hepatitis. Additionally, a drug called methisazone is reported by some to show a short-term protective effect against smallpox, and another called amantidine (sold under the name *Symmetrel*) appears to be of some value in the prevention of the Asian flu.

In regard to the actual treatment of viral infections, antibiotics, sulfa drugs, and other antimicrobial agents are, with but few exceptions, without effect. (They are of value, of course, when a bacterial invasion follows in the wake of a viral attack.) An outstanding exception is the drug idoxuridine, which gives good results in treating eye infections caused by the cold-sore virus. Antibiotics are effective in the treatment of psittacosis (parrot fever) and trachoma, but most virologists no longer consider the causative agents of these infections to be viruses.

The diagnosis of a viral infection centers upon the signs and symptoms and a variety of laboratory tests, many of which are highly sophisticated. Some infections are so characteristic that the layman himself often "makes the diagnosis." We are reminded of measles, mumps, and chickenpox. But an error can be committed, even by the doctor, which all goes to underscore the point that a viral infection can never be diagnosed with absolute certainty unless the virus or its telltale antibody is demonstrated. There are one or two infections, however, where the inclusion bodies present in the affected cells are sufficiently characteristic to afford a definitive diagnosis.

In the main, laboratory tests are based upon the fact that a virus can be used to identify an antibody and vice versa. For example, if much diluted blood serum is added to a *known* cytopathic virus and the resulting mixture fails to destroy the cells of a tissue culture, this shows conclusively that the blood contains the neutralizing antibody against this particular virus—the very same virus that has stimulated the patient's plasma cells. All other antibodies would fail to neutralize the virus and the cells would be destroyed (a negative test). Again, many viruses cause animal red cells to clump together, a phenomenon called hemagglutination. Thus, if such a known virus when mixed with the patient's blood serum is neutralized and thereby prevented from bringing about this action, it shows that the infection in question is caused by the test virus; contrariwise, hemagglutination constitutes a negative test.

And then there is the revolutionary fluorescent antibody technique. In this test, known antibody is combined ("tagged") with the fluorescent dye, fluorescein, and added to a smear or tissue culture of cells derived from the patient. After a few minutes the antibody is washed away and the preparation examined microscopically under ultraviolet light. The presence of luminous areas within the cell is visible proof that the antibody has combined with the corresponding virus. For example, the fluorescent-tagged rabies antibody will produce luminous areas only when the cells contain rabies virus.

7

Lady Montagu

IT STARTED OUT SOMETHING LIKE THE FLU— TERRIBLE HEADache, chills, aching back and limbs, elevated temperature, and all the rest. This was the picture in the morning and by late that evening the patient was delirious and unable to get up off his back. The face was flushed, the pulse bouncing, and the thermometer read 107° F.!

The following day it was more of the same, notwithstanding huge doses of antibiotics. And then on the third day a raised rash appeared on the face that rapidly spread to the trunk and arms and legs. The doctor suspected a strange case of measles or scarlet fever, a hypothesis he had to discard two days later when the rash gave way to a rampant eruption of blisters—blisters even on the palms, the soles, and the lining of the mouth and throat.

On the sixth day—the day the patient said he felt a little better—the blisters were noted to be filling out with pus and changing to a greenish-yellow color. And then they began to itch terribly, and by the following day the patient was about out of his mind. The doctor now realized that these sores were in all likelihood the pustules—or "pocks"—of smallpox. The local health authorities imposed a strict quarantine and all exposed persons were "ordered" vaccinated or revaccinated. Meanwhile, the state laboratory had demonstrated both inclusion bodies (Guarnieri bodies) in the cells of a pock specimen and antibodies against the smallpox virus in a blood specimen. Happily, no other cases developed and the patient himself went on to recovery, the pocks having dried up and fallen off by two weeks' time. One or two pock marks were to remain, emblems of the fact that this particular individual would probably never have the disease again.

The smallpox virus is unusually resistant to drying, and the live virus can be easily isolated from scabs, floor sweepings, and clothing. It is transmitted from man to man by contact with the patient or his immediate surroundings, usually entering the body proper via the nose or mouth. During the incubation period, which runs about twelve days, the virus rapidly multiplies in the lymph nodes and related tissues and from these sites enters the blood to attack the body at large. Once the signs and symptoms appear there is nothing of a specific nature the doctor can do.

Smallpox, or variola, is an ancient disease that flourished in epidemic form in China as early as the twelfth century B.C. It spread throughout Europe during the Crusades and reached the Western Hemisphere sometime in the early sixteenth century. The slave trade in particular did much to bolster the disease in the early history of America. Today smallpox occurs chiefly in Africa and Asia, especially India and Pakistan, and its introduction into America from these areas remains a distinct possibility. But the United States has been fantastically fortunate in this Jet Age, for whereas the year 1921 yielded 110,000 cases of smallpox, there has not been a confirmed case of this disease in this country since 1949.

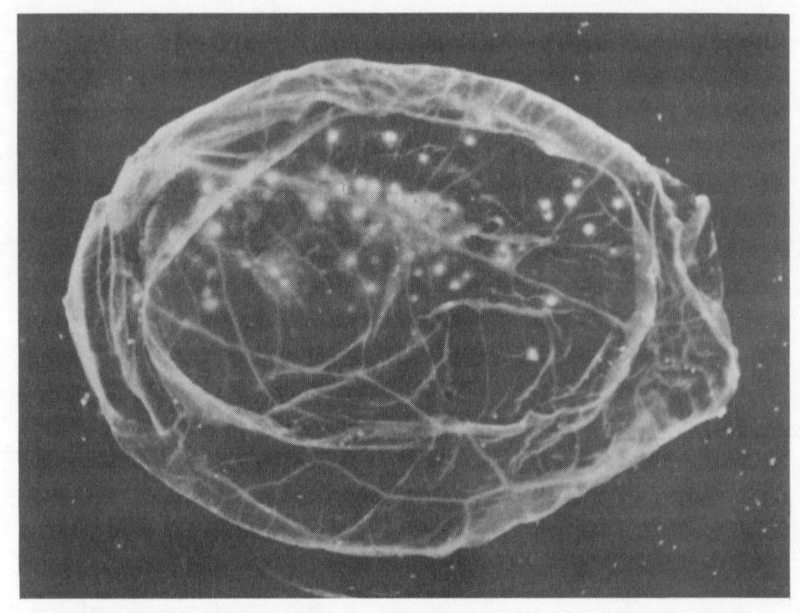

The effect of the smallpox virus on the chorio-allantoic membrane of a fertile hen's egg after 72 hours' incubation. Each white pocklike lesion represents the destruction set off by a single virion. (From Hahon and Ratner, *Journal of Bacteriology*, 1957.)

A person's chances of pulling through an attack of smallpox relate in the main to which of the two forms of the disease is involved—variola minor or variola major. Although the causative viruses are indistinguishable, the "minor" is marked by a fatality rate of about one percent compared to a "major" fatality rate ranging from 20 to 40 percent. By way of example, of 35,152 cases of variola minor in Minnesota in 1923, there were only 108 deaths;

the following year in the same state there were 365 deaths among 1,430 cases of variola major.

The answer to smallpox resides solely in vaccination, a practice that dates back, in principle at least, to ancient times. (Antibiotics and sulfas are useful only in the control of bacterial complications.) The custom in the Orient was the application of dried crusts of smallpox pustules to either the skin or nasal membranes or the actual ingestion of such crusts. The infection produced by this practice (called variolation) was usually much less severe than naturally acquired smallpox and the fatality rate considerably less than that of the natural disease. Above all, the individual was rendered immune to natural smallpox. Lady Mary Wortley Montagu, wife of the British ambassador at Constantinople, was greatly impressed by this oriental magic and is credited with introducing it into Europe in 1718.

But variolation proved to hold the uncertain seeds of grave danger and most people understandably eschewed those ugly dried crusts. Besides, another kind of magic was coming to the fore, one much more interesting. Persons who contracted cowpox from infected cattle it was noted were immune to smallpox. Moreover, cowpox proved to be an exceedingly mild disease in man, so mild that it was confined to the site of contact. Milkmaids, for instance, customarily developed only a pustule or two on the hand, and yet they became immune—immune to cowpox and immune to smallpox.

And then in one of the great moments in medicine, Edward Jenner, an English country doctor, turned this cowpox phenomenon into an artificial means of immunization. In 1796, using material taken from a cowpox pustule on the hand of a young milkmaid, Jenner inoculated a small boy and six weeks later applied the acid test by inoculating the boy with pus of the terrible smallpox. The boy failed to develop the disease! Because of the role the cow played in all this, the material Jenner used to provoke the immunity was called a "vaccine" and the procedure itself "vaccination," after *vaccus* (Latin for cow). Today both terms are employed in all instances of active artificial immunization.

The modern smallpox vaccine is composed of vaccinia virus, a man-made derivative of the cowpox virus. In the usual mode of manufacture, seed vaccinia virus is inoculated into the skin of calves and then harvested by extracting the watery fluid (lymph) which forms within the resulting blisters, or vesicles. To this glycerin is added and the mixture emulsified in a blender to form the finished product. When stored at below freezing temperatures the vaccine maintains its potency for a very long time; just at the freezing point it is good for about six months. A freeze-dried vaccine is also available that maintains its potency even in tropical heat. Vaccinia virus may also be grown in fertile hens' eggs and chick-embryo tissue cultures.

Vaccination is conventionally carried out by the so-called multiple-pressure method, whereby a needle held at a 45-degree angle to the skin surface is moved up and down through a drop of vaccine with pressure on the skin sufficient to slightly puncture it. (For the rapid vaccination of large groups a jet injection apparatus is used.) In the previously nonimmunized, a small, circumscribed, solid elevation of the skin (called a papule) appears on the third or fourth day, which becomes vesicular (or blister-like) on the sixth or seventh day and pustular (pus-filled) on the eleventh or twelfth day. The pustule—or pock—now

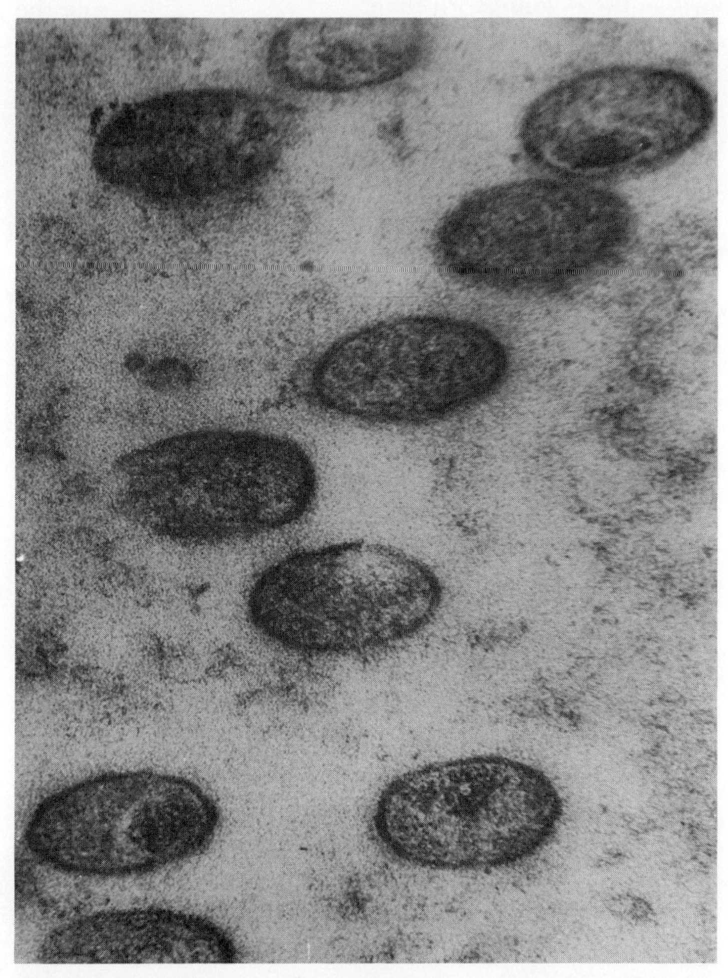

The vaccinia virus magnified 180,000 times. (Courtesy Councilman Morgan, M.D. and the Department of Microbiology, College of Physicians and Surgeons of Columbia University.)

regresses, resulting in a crust that falls off about three weeks after inoculation. This leaves a red pitted scar which turns white with time.

In persons partially immune to smallpox as a result of having been vaccinated some years previously, the foregoing sequence of events is almost always telescoped into a period of about a week. This is the so-called accelerated or vaccinoid reaction. In persons fully immune to smallpox as a result of having been vaccinated a few month previously, only a red wheal develops at the site of inoculation; that is, no pustule forms and no scar results. A person who shows *no* response must *not* be considered immune—*as is generally supposed*—and should be revaccinated promptly. Either the vaccine was poor or the technique was faulty.

The most common complication of vaccination is a bacterial infection of the pustule as a result of neglect, scratching, or skin infection elsewhere. Very rare complications are generalized vaccinia (a condition marked by fever and eruptions other than at the site of the inoculation) and inflammation of the brain (postvaccinal encephalitis). Also, there are certain contraindications, including eczema, leukemia, pregnancy and agammaglobulinemia (a condition in which the blood is without antibodies of any kind). Those with eczema should not even come into contact with recently vaccinated persons. The vaccinia virus readily attacks areas of the skin subject to eczema and sometimes even healthy skin. The fatality rate among infants with "eczema vaccinatum" may run as high as 30 percent.

A comprehensive national survey to ascertain the frequency of complications associated with vaccination in the United States during 1963 showed that among more than 6 million primary vaccinees and nearly 8 million revaccinees there were 12 cases of encephalitis and 108 cases of eczema vaccinatum. Seven persons died. A substantial number of less serious complications, some of which resulted in hospitalization, were also reported.

Vaccination, then, is associated with a definite, measurable risk of morbidity—and, rarely, death—and not a few doctors nowadays are having second thoughts about it. Be that as it may, the reservoirs of endemic smallpox have changed very little over the years whereas international travel has increased substantially, and until such reservoirs are eradicated vaccination clearly represents the only currently practicable approach for community protection in the United States.

The Public Health Service Advisory Committee on Immunization Practices recommends primary vaccination between the first and second birthdays and at any age under conditions of exposure or foreign travel, and revaccination at three-year intervals for persons who conceivably might be exposed in endemic or potentially endemic areas by virtue of travel and for persons likely to be exposed by newly introduced infection into the country, particularly hospital personnel, public health workers and morticians. All other persons, the committee concludes, should be vaccinated at approximately ten-year intervals.

8

Two For One

THE BOY DID NOT FEEL QUITE UP TO GOING TO SCHOOL, AND he had a slight temperature. This was in the morning, and by early afternoon his mother noted the beginnings of a rash, one that gave way in a couple of days to fragile dewdrop-like blisters encircled by a red halo; in turn, these developed into sores which looked just like the textbook pustules, or pocks, of smallpox. But the boy had been vaccinated against that disease, and besides, his mother knew from her experience with his siblings that this was a full-blown case of good old-fashioned chickenpox.

Of course, the doctor was consulted to "confirm the diagnosis." Meanwhile, the involvement had spread and by the fifth day the young patient appeared to be enveloped in spots, blisters, and pocks. The itching was unbearable and in his sleep he scratched so much that some of the lesions became infected. An antihistamine gave some relief and, most important, the secondary infection yielded readily to an antibiotic. The rest was, as the doctors say, "uneventful," and the boy went on to complete recovery. Above all, he would probably never have chickenpox again.

Chickenpox, or varicella, may strike at any age, but it is primarily a disease of childhood, with most cases occurring between the ages of two and eight years. For the most part, it arrives during the cooler months and in epidemic proportions. Interestingly, by adulthood most persons are immune to the disease whether or not they have ever had chickenpox as a child. Though it is well known to be among the mildest of infections, the truth of the matter is that chickenpox can prove deadly, especially in adults. Pneumonia is always a possibility and so too, although very rare, is encephalitis. Recently, two deaths from chickenpox pneumonia occurred in the same family, a nine-year-old boy and a six-year-old girl. In infants receiving cortisone-like drugs and in those with leukemia chickenpox may have a fatal outcome. Thus, all cases must be considered potentially dangerous and handled with utmost care. As of this date a satisfactory vaccine against the disease has not been developed.

The chickenpox virus is brick-shaped and very large as viruses go. Apparently, man is its only victim. All attempts to grow the virus failed until Thomas Weller in 1953 succeeded in culturing it in human tissue. The virus is present in the fluid of the blisters and from here travels to the respiratory tract of the next host by airborne means. It then invades the blood and eventually the cells of the skin, the chief target area; by two weeks' time the signs and symptoms are manifest. Scrapings of young blisters, specially stained, disclose highly characteristic giant multinuclear cells containing reddish inclusion bodies within the nucleus, a finding that will serve to differentiate a case of chickenpox from a mild or modified case of smallpox. As indicated, however, chickenpox is almost always diagnosed on a basis of the clinical picture, especially when the patient has either had smallpox or has been vaccinated against it.

For a long time it has been clear that there is a close relationship between the virus of chickenpox and the virus of shingles ("herpes zoster") and now, lo and behold, the etiologic agents of both infections turn out to be the one and the same virus! A person can contract a case of chickenpox from another person with shingles and, apparently, vice versa. The laboratory has demonstrated conclusively that the viruses are identical *in vitro* ("in the test tube"). There is considerable difference of opinion among authorities as to the mechanisms and explanations behind this strange and interesting situation, but the general feeling seems to be that the first invasion of the body by the so-called VZ (varicella-zoster) virus results in chickenpox, whereas shingles results from either the reinvasion or the activation of a latent virus. Support for the "latent theory" stems from the fact that patients with shingles can produce no history of prior contact with an external source of the AZ virus and from the fact that certain stimuli and conditions are known to trigger a case of the infection—injury, drugs, leukemia, and hormones, for example.

Shingles is primarily a disease of adults (with the highest prevalence somewhere between the ages of fifty and sixty), but children are afflicted and some cases occur even in infants. In the wake of an incubation period of about two weeks, there is pain and burning and a prickling feeling in the skin, followed in a few days by a rash and then blisters very much like those of chickenpox. In contrast to chickenpox, however, the eruption develops into a band-like pattern, characteristically on one side and usually on the chest. The explanation for this is that the virus preferentially attacks the spinal nerves, following them to their termination at the body surface. There is an intense burning pain in the involved area and in some cases malaise, headache, and fever. Shingles usually runs for about two weeks, but some patients suffer for much longer than this. Indeed, the pain sometimes persists even after the eruption has gone and may demand neurosurgery. Again, if the virus has attacked the cranial nerves of the head the eyes may become involved, so involved, in fact, that there may be scarring of the cornea.

For the present, anyway, there is, as in chickenpox, neither a mode of prevention nor a specific remedy available. About the only thing one can do is to keep his fingers crossed that nothing will happen to stir up his VZ virus. Too, perhaps grandparents should stay away from their AZ-afflicted grandchildren.

Closely related to the VZ virus is the causative agent of herpes simplex—the "cold sore virus." The vast majority of us harbor the virus, usually in the nose and mouth, and have harbored it since the age of three or four. Nine times out of ten the first exposure to the virus merely results in an inapparent infection—no signs, no symptoms; in short, the herpes simplex virus invades the cells and stays there, perhaps for life. In those cases where the infection does become manifest upon first exposure, the usual situation is one of fever, malaise, and vesicular eruptions of the skin and mucous membranes of the nose and mouth. Indeed, there may be small outbreaks of the infection in prisons, orphanages, and among hospital personnel. Severe cases of herpes simplex may involve the gums, throat, eyes, genitals, liver, spleen, lungs, kidneys, and brain. Mild or severe, the disease usually runs its course in one to three weeks.

In response to the invasion, whether manifest or not, the plasma cells respond by pouring forth antibody which apparently neutralizes the virus in the blood and other

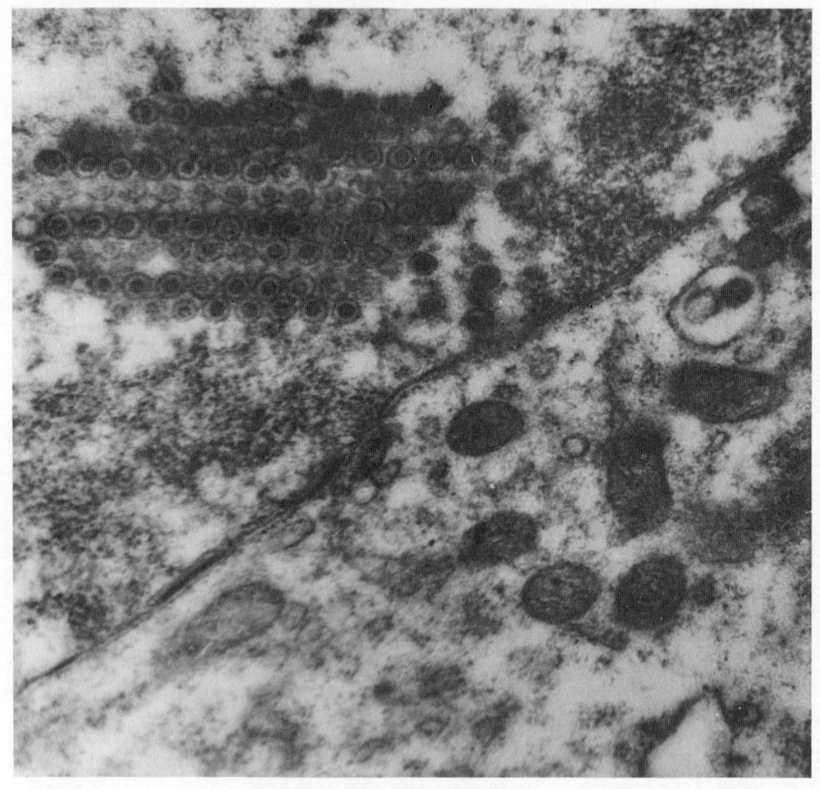

The cold sore virus crystallized within the nucleus of a cell. The nuclear membrane traverses the photo from lower left to upper right. Note that each virion of the crystal has a dense central body and a single surrounding membrane. Magnified 76,000 times. (Courtesy Councilman Morgan, M.D. and The Department of Microbiology, College of Physicians and Surgeons of Columbia University.)

body fluids but fails to eradicate the virus in such focal areas as the cells of the nose and mouth. And this is important to remember because when the resident virus acts up, as it is prone to do, the trouble is generally confined (by antibody) to the original portal of entry. In other words, those irksome fever blisters or cold sores are caused by an out-of-hand herpes virus that can probably trace its ancestry back to our early childhood.

Persons vary greatly in their susceptibility to recurrent cold sores, a fact which underscores the interplay between host and parasite in the complex context of third-party external factors. More to the point, the following "stimuli," among others, are known to provoke cold sores: local initiation, drafts of cold air, heat, light, fever, respiratory infections, upset stomach, injury, neuralgia, physical exertion, pregnancy, menstruation, and emotional stress. Perhaps it is scientifically accurate to say that "anything" can set off an attack.

Unhappily, nothing can be done to prevent herpes simplex, at least for the present, and for the most part little can be done in the way of specific therapy except in instances of the eye. Here the antiviral agent, idoxuridine, may ameliorate the acute manifestations and prevent damage to the cornea. Perhaps one day the chemist will come up with a drug totally effective against all facets of the herpes simplex virus.

9

Spots

OUR YOUNG FRIEND FROM CHAPTER 8 ONCE AGAIN DID NOT feel well. He was sickish and feverish and he was sneezing and coughing. His nose was running profusely and his eyes were red and tearing and very sensitive to light. On the second day the doctor noted tiny white spots inside the mouth (adjacent to the molar teeth) and on the third day a reddish brown rash appeared behind the ears. By the fourth day the rash had spread from head to toe and the boy was now very ill. The temperature rose to 103° F. and his coughing could be heard throughout the house. The stethoscope disclosed lung involvement and the flashlight illuminated the beginnings of a middle-ear infection. Antibiotics were given around the clock to control these rather ominous developments, and aspirin was ordered to bring down the temperature and help relieve the aches and pains. In a day's time these medicaments did indeed have the desired effect and, too, the rash upon the face had begun to disappear. Everything now went along uneventfully and the patient came through with a lifelong supply of antibodies against the virus of measles.

Measles (or rubeola, or morbilli) is cosmopolitan in its distribution and one of the most contagious diseases

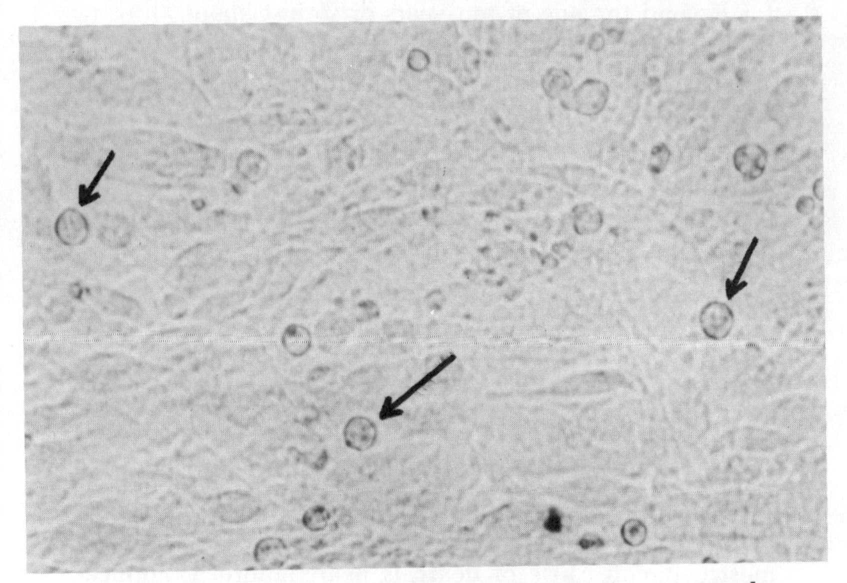

Photomicrograph shows the effect of the measles virus on the corneal cells of the eye of a rabbit. The normal, more or less rectangular cells become spherical and darker. Arrows indicate three damaged cells. (Courtesy of the National Medical Audiovisual Center, Atlanta, Georgia.)

known. The incidence in this country runs around a half million cases a year, to use a conservative statistic, with most outbreaks occurring in late winter and early spring. Measles recurs in epidemic cycles at two- and three-year intervals, and this is thought to result from the introduction of new susceptibles into the population by birth and ingress from other areas. Further, it is believed that the virus is introduced into the involved population from external sources, no doubt by incoming susceptibles. Like chickenpox, measles is essentially a disease of childhood, for beyond the age of ten years or so just about all of us have specific antibodies against the infection.

Measles is acquired and disseminated via the respiratory tract, one of the chief target areas in addition to the skin. The virus proliferates in the mucosa that lines the tract, eventually making its way into the blood and thence the body at large. The incubation period averages about eleven days. The disease is always to be taken seriously because the weakened respiratory tract can easily give way to a severe and dangerous bacterial pneumonia. Also, middle-ear involvement is not uncommon and sometimes not without grave repercussions. Moreover, though rare, there is always the possibility of encephalitis, a development that carries a fatality rate of about 25 percent. And of those who do pull through there are residual effects in almost half of the cases. In patients fifty years of age or older the most common cause of death is heart failure. Complications are of all sorts and the highest fatality rates are in children under five and in the aged. In what may come as a surprise to many, in 1956 measles was responsible for over twice as many deaths as polio in children under fifteen years of age.

Perhaps the most insidious feature of the infection was uncovered in 1969 by John L. Sever of the National Institute of Neurological Diseases and Wolfgang Zeman of Indiana University Medical Center. These researchers and their associates isolated the measles virus from the brain tissue of youthful victims of a strange disease labeled subacute sclerosing panencephalitis. Commonly called SSPE, the malady is believed to affect 100 to 200 American children annually. It is nearly always fatal within a year. The first inkling of trouble comes when children start doing poorly in school or develop personality changes, followed in rather short order by such overt signs as falling, stumbling, jerking, and convulsions. Terminally, the victim passes into a coma. Sever and Zeman and others believe that what we have here is a smoldering virus. Moreover, there is now the revolutionary idea that a host of still-unsolved maladies may be smoldering holdovers from other viral infections. Cancer, multiple sclerosis, and Parkinson's disease are already being talked about in this light.

A diagnosis of measles is usually made by the parent hours before the doctor arrives—and the diagnosis is usually correct. Any feverish, blotched-faced, bleary-eyed child who breathes open-mouthed between paroxysms of sneezing and coughing almost always has measles. And the presence of those tiny white spots on the inside of the mouth—named Koplik's spots after their discoverer Henry Koplik—is almost tantamount to a positive laboratory diagnosis. This is especially true when there is known measles in the community. Nonetheless a misdiagnosis is possible on a basis of the clinical picture alone. Among other mimickers are German measles, scarlet fever, infectious mononucleosis, syphilis, and infections due to Coxsackie and ECHO viruses. An unequivocal diagnosis, as in all infections, of course, is made only in the microbiological laboratory. This entails either the isolation of the virus from an appropriate specimen or else the demonstration of a steady rise in antibody concentration. Additionally, stained smears of nasal secretions of measles patients disclose so-called Worthin-Finkeldey cells (huge multinucleated cells).

As in all viral diseases the central consideration revolves around immunity. Infants born of immune mothers have received a big dose of antibody *in utero* and are immune for about six months. An actual attack of the disease, as already noted, provokes an immunity that almost always is good for life. The classic exception would be in persons whose plasma cells fail to produce antibody of any kind (a condition called agammaglobulinemia). Measles may actually be prevented in exposed susceptible persons by giving them a shot of gamma globulin prepared from immune blood, and this tactic is employed in those cases where an attack of the disease could be disastrous to an already ill or debilitated patient. However, as in the case of the newborn this is passive immunity and good for only a short time.

For a truly effective active immunity—without having to suffer the disease—a live, attenuated vaccine is now available. Attenuated by multiple passage of the virus in monkey, human, or chick-embryo cell cultures, the vaccine stimulates a protective concentration of antibody in two to three weeks following vaccination. The protection is believed to be good for life. In 1966 when the vaccine came into general use, there were 204,136 reports of measles and in 1968, only two years later, this figure was down to only 22,538 cases. Thus, there is excellent reason to believe that the infection is destined for extinction.

10

Germany

IN THE EARLY PART OF THE EIGHTEENTH CENTURY AN ENGLISH physician described a very mild disease marked by a rash much like that of measles and scarlet fever. Years later epidemics of the same illness were recognized in Germany, where it became known as *Rötheln* (German for redness). But the English-speaking world found *Rötheln* too harsh and too foreign and in 1866 another English doctor proposed the name rubella, from the Latin *rubellus* (reddish). And as an etymological finale, a New York pediatrician by the name of J. Lewis Smith in 1874 referred to an outbreak of rubella as German measles.

From the beginning of the present century there was general agreement on the viral etiology of German measles, but it was not until 1961 that the virus was isolated and identified as the cause of the disease. Unlike most viruses, the rubella agent does not kill the cells in which it multiplies (the cytopathic effect) and therefore does not make its presence directly known in tissue cultures; hence this historical tardiness. The ingenious approach that finally bore fruit—known as the "interference technique"—runs as follows: A sample of an *unknown* noncytopathic virus is added to a medium containing German measles antibody

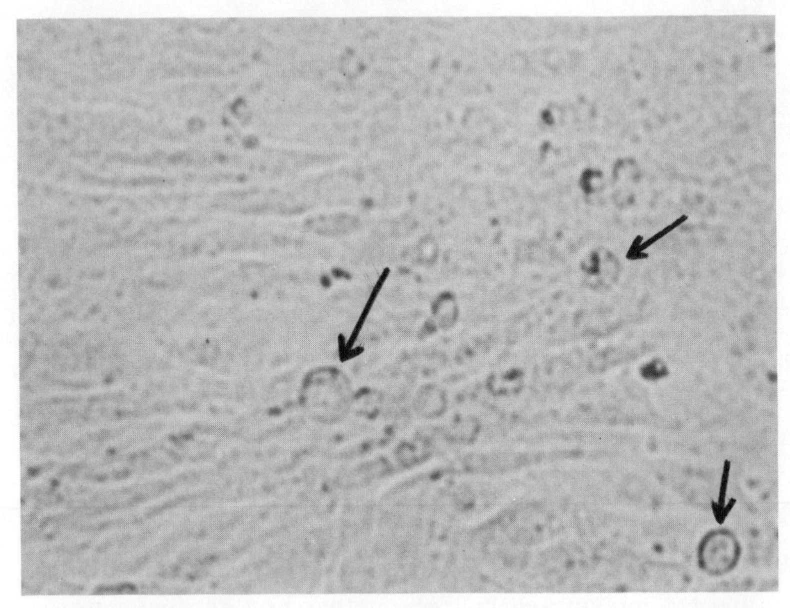

Photomicrograph shows the effect of the German measles virus on the corneal cells of the eye of a rabbit. The somewhat rectangular cells become spherical and darker. (Courtesy of the National Medical Audiovisual Center, Atlanta Georgia.)

and the mixture is then inoculated into a tissue culture of fresh monkey kidney cells. If the unknown is indeed the German measles virus, the antibody neutralizes it and thereby prevents it from provoking the cells to form the virus-inhibiting agent interferon. Thus, the challenge cytopathic ECHO virus that is now added can exercise its *visible* cytopathic effects. Contrariwise, if the virus under study is *not* that of German measles, the antibody has *no* effect and it (the virus) enters the cells and *provokes* interferon—which then "interferes" with the challenge

ECHO virus and prevents the latter from destroying the cells. In sum, cytopathic action signals the *presence* of the German measles virus, and no cytopathic action signals the *absence* of the German measles virus.

German measles is primarily an infection of childhood characterized by enlarged lymph nodes, usually those behind the ears or at the back of the head just above the hairline, and a rash which is quite changeable and difficult to distinguish from that of other diseases. The virus is spread from person to person through contact or "nasal spray," and once it has gained entrance into the body the signs and symptoms of the infection appear in about two weeks. German measles is exceedingly mild and only on rare occasions does it result in serious complications, the most important of which is encephalitis. Not uncommonly the infection may be present without any apparent symptoms at all.

And this was the sum and substance of the matter until 1941, the year in which an Australian ophthalmologist by the name of N. McAlister Gregg startled the medical world by the discovery that congenital cataract is a common occurrence when German measles strikes the mother during the first three months of pregnancy. In addition to cataract, a condition in which opaque areas form in an otherwise clear lens of the eye, Dr. Gregg's classic paper noted the low birth weight, congenital heart defects, and high mortality in such infants. Historically, this established the German measles virus as the first clearly defined agent to cause developmental abnormalities—a teratogen (as such agents are called) to be joined in years to come by another "German product" called Thalidomide. The precise mechanism whereby the virus produces these abnormalities is not known, but it can be shown in the laboratory that the virus inhibits the division of embryo cells in tissue culture and also does visible damage to the chromosomes.

The full impact of Gregg's findings came in 1964 when, in the winter and spring of that year, one of the most severe epidemics in history swept across the United States, leaving in its wake an estimated 20,000 congenitally defective babies. In addition to heart disease and cataracts, the most commonly noted abnormalities during the first month after birth, there were deafness, brain damage, hepatitis, bone lesions, enlarged spleen, enlarged lymph nodes, low birth weight, hepatitis, and purpura (a bleeding condition that causes numerous spots, like tiny bruises, in the skin and mucous membranes). These defects occurred both singly and in all sorts of combinations, and surprisingly the virus was found to persist in afflicted babies for months in the face of high levels of antibody in the blood. This cannot be explained and serves as just another reminder of how much more there is to learn about the forces and mechanisms of immunity.

The persistence of the virus in afflicted babies amounts to a real hazard to hospital personnel, particularly since some of these babies appear to be normal. And even outside the hospital there is grave danger to women in early pregnancy because the German measles virus is present in the throat for a week or more before the rash appears and for a week or more after it disappears. Thus, the hapless host is often unaware of exposure until the infection becomes manifest. There is a good fifty-fifty chance of damage in the first month of pregnancy. This drops to about 25 percent in the second month and 10 percent in the third. Many hospitals in this country, on the basis of these figures, will permit a therapeutic abortion if German measles occurs within three months of pregnancy.

Recent reports show, however, that the third month of pregnancy by no means concludes the danger. More subtle but in many instances no less serious handicaps can occur after this time. In one study involving 24 women with clinical evidence of the infection between the 14th and 31st weeks of gestation there were two fetal losses and 15 live-born children displaying an array of abnormalities ranging from mental retardation to an enlarged spleen.

An attack of German measles almost always confers a lifelong immunity and this would indicate that a solution to the disease resides in a vaccine made from the virus. So it does, and a vaccine is now about to be put into general use. The development of this particular vaccine proved especially irksome and difficult because on the one hand the killed virus does not provoke sufficient antibody, and on the other hand the live virus can easily render the vaccinated person infective to others—and even the mildest case can result in severe repercussions in pregnancy. Somewhat paradoxically, in other infections the dissemination of a virus by vaccinated persons is a desirable feature in that it confers widespread immunity without danger. The trick, then, was to tame the German measles virus to such an extent that it could not possibly be spread to susceptible contacts, and in 1965 Paul D. Parkman and Harry Meyer, Jr. did this very thing by a succession of culturings in monkey kidney cells. Based on present experience and information rubella may one day join hands with rubeola and walk hand in hand into well-deserved oblivion.

11

Beside The Ear

NINE-YEAR-OLD BILLY HAD A TEMPERATURE, CHILLS, HEADache, sickish feeling, and loss of appetite. This was on Sunday, and on Monday it was a little more of the same. And then on Tuesday his mother noted swelling about the angle of the right jaw, and it was so painful that Billy could hardly open his mouth. The swelling advanced a little on Wednesday, but by Thursday it began to subside and, to abbreviate the story, by Sunday Billy was all set to return to school the following morning.

Billy's illness, to be sure, was mumps, or, more formally, parotitis (inflammation of the parotid gland). Parotid in Greek means "beside the ear" and this is pretty much its location. In all, there are three pairs of salivary glands: the parotids, the submaxillaries (below the lower jaws), and the sublinguals (beneath the tongue on either side). These other glands may or may not be involved, and even when they are the swelling is seldom noticeable.

Ernest W. Goodpasture and his co-workers isolated the mumps virus in 1934 by inoculating saliva of patients into the parotid gland of monkeys. In addition to causing mumps in that animal, the virus may also incite a fatal encephalitis in mice and hamsters. The mumps virus is a member of the myxovirus family, which also includes, we shall recall, the viruses of influenza. As presently understood, the virus enters the host by the nose or mouth (through infected droplets or direct contact) and then makes its way into the blood to be carried far and wide to other tissues. The incubation period usually runs between two and three weeks, but may go for a month or more.

The mumps virus is often present after the signs and symptoms disappear, and, further, it may be recovered

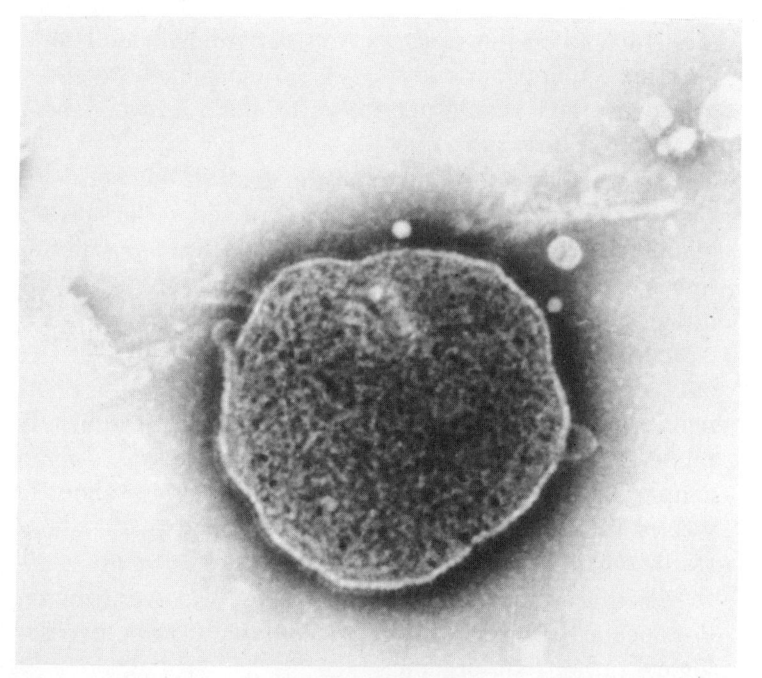

The mumps virus—showing a single virion—magnified 124,800 times. (Courtesy, Department of Virus and Cell Biology, Merck Sharp and Dohme Research Laboratories, West Point, Pennsylvania.)

from patients with *inapparent* infection. Although communicability is comparatively low, high attack rates are noted in all age groups in such isolated populations as the Pacific Islands. In all other areas most cases occur between the ages of five and fifteen. In contrast to measles and chickenpox, susceptibility to mumps occurs in many persons reaching adulthood, a finding that underscores the fact that infection, either apparent or inapparent, is needed to stimulate a protective level of antibody. But once stimulated, the immunity conferred is generally of long duration, with second and third attacks being very rare. Legend to the contrary, the patient with unilateral parotitis (mumps on one side) is no more likely to suffer a second attack than a patient with bilateral involvement (mumps on both sides).

Fever, malaise, loss of appetite, and so on, correctly suggest that mumps is by no means confined to the salivary glands alone. The most common extrasalivary manifestation is orchitis, or inflammation of the testis. Though infrequent before puberty, the overall incidence in the adult is close to 25 percent of all cases. The affected testis (usually only one is involved) may enlarge two or three times normal size, is painful, and, as the doctors say, "exquisitely tender." The situation characteristically progresses for a couple of days or so and then subsides in about a week's time. Much to-do has always been made about sterility in the wake of mumps orchitis, but on good authority this development seldom ensues. And even when it does occur the underlying factor is thought to be psychological rather than organic.

Not uncommonly, too, the mumps virus noticeably affects the central nervous system causing, in severe cases, meningoencephalitis (inflammation of the brain and the meninges, or membranous wrappings about the brain). In the latter event there is fever, headache, stiff neck, nausea, vomiting, and drowsiness. And of some diagnostic importance, the eyes are extremely sensitive to light (photophobia) and the cerebrospinal fluid (the fluid that bathes the brain and spinal cord) is elevated in both pressure and cell count. Death may occur, but the usual outcome is a relatively slow convalescence without aftereffects, or "sequelae."

The mumps virus may also attack the pancreas, heart, thyroid gland, breasts, tear glands, ears, kidneys, and ovaries. Inflammation of the ovary (oophoritis) is thought to be fairly common but is rarely thought to result in sterility. According to expert opinion oophoritis should be considered when the patient complains of pain in the lower abdomen or lower back and an enlarged ovary can be felt upon examination.

The diagnosis of mumps on a basis of signs and symptoms generally presents little difficulty, especially when the patient is known to have been exposed to the disease. Moreover, tests are now available that can pinpoint mumps antibody in a blood specimen and thereby afford a definitive diagnosis. The standard test of this character uses two specimens, the first taken as early as possible and the second about a week after onset of clinical manifestations. A positive test is signaled by a *sharp increase* in antibody, an event that can only occur during an actual attack. A negative test is constituted by a steady level of antibody as well as no antibody at all. (A steady level means that a mumps infection occurred sometime in the past.) Recently, a test was developed that zeros in on a special kind of mumps antibody, an antibody which appears during an attack and disappears after the attack.

Put another way, the presence of this special antibody in a blood specimen constitutes a positive test. Since only one blood specimen is needed, a definitive diagnosis can be made in a matter of minutes.

In regard to prevention, gamma globulin (derived from blood known to contain mumps antibody) may prevent the disease when given during the incubation period. As we know, however, this kind of immunity is good only for a matter of weeks and is therefore applicable only to treat children already ill or to prevent the infection on short notice in those with a condition mumps would compound and exacerbate. A live, attenuated vaccine is now available that shows much promise. Although it is still not known whether the vaccine will provide protection after exposure to mumps, its ability to stimulate detectable antibodies in 95 percent of susceptible vaccinees has prompted the Public Health Service to recommend its use in all susceptible children over one year of age. It may well be that the involvement "beside the ear" may before long become a thing of the past.

12

Blood and Oysters

THE SOLDIER WAS IN BAD SHAPE: HEADACHE, FEVER, ABDOMinal pain, diarrhea, and nausea and vomiting. Actually, he had not felt himself for sometime. His weight was down from not eating and, interestingly, he had developed a strong distaste for cigarettes. On the day the patient entered the hospital he passed dark urine and on the following day his body had turned yellow from head to toe—even the eyes and inside the mouth were yellow. This yellowing of the skin, plus an enlarged liver and high bile content of the blood, was jaundice pure and simple. But what kind of jaundice? The medical dictionary listed more than forty types, none of which, to shorten a long story, applied to this particular case.

The patient was placed on a bland diet and given glucose and salt solutions intravenously to bolster the caloric intake and replace the fluid lost through vomiting and diarrhea. The acute signs began to remit "on their own" soon after the yellowing appeared, and the latter itself cleared up in about a week. But for three weeks the patient was not strong enough to get out of bed, and even when he was able to get up the slightest exertion proved tiring and sometimes overwhelming.

This particular case was typical of a new and mysterious illness that struck 50,000 of our troops in the early days of World War II. Fortunately, it did not remain a complete mystery for long because some excellent detective work turned up a common denominator among the victims of the epidemic. They had all been inoculated with a yellow-fever vaccine containing human blood serum, a serum which in turn contained a virus capable of causing the disease just described. The disease was quickly labeled "serum hepatitis" (inflammation of the liver) and the virus, one of the smallest ever to be discovered, was dubbed "SH" (after serum hepatitis).

Meanwhile, another viral mystery was giving ground—an infection identical to serum hepatitis but caused by a different virus. To this disease was given the name "infectious hepatitis" and to the virus was given the designation "IH." Commonly, the SH virus and IH virus are also referred to as virus A and virus B, respectively. The two viruses are closely related, attack man only, and possess a most unusual predilection for the liver. And it is the attack on that organ which explains the disease *in toto*. For years these viruses refused to be cultured, but today can be grown successfully in human cell cultures.

The IH virus resides in the intestinal tract and appears in the feces, meaning that it is easily spread under conditions of overcrowding and poor sanitation. Explosive outbreaks generally stem from contaminated food and water. Oysters, clams, and other shellfish have proved particularly dangerous because they tend to concentrate the virus, which is ever present in polluted tidal waters. No one pretends to know the exact incidence of the disease because there is good reason to believe that most cases go unrecognized. Those which are recognized probably average about 150,000 a year, and the figure appears to be rising.

The SH virus is always associated with human blood, and this covers a lot of ground when we consider that one-millionth of a drop of contaminated blood is all that is needed to set off an infection. Blood transfusions come first to mind, and recent guesses have it that every hundredth bottle of blood contains the SH virus. To date, no foolproof method is available to prevent this. Improperly sterilized hypodermic needles, syringes, intravenous tubing, and the like are also potential hazards, and doctors, nurses, and dentists are continually being urged to use disposable devices. Among medical personnel and laboratory workers serum hepatitis amounts to an occupational hazard, and the disease is thought to be common among addicts who share the same syringe and among those who go to tattoo parlors.

Other key differences between infectious and serum hepatitis, in addition to the virus reservoir, relate to the incubation period, immunity, and mortality rate. The incubation period for infectious hepatitis is from two to six weeks and for serum hepatitis from two to six *months*. Both viruses are thought to provoke a fair resistance to subsequent encounters, with expert opinion giving the edge to IH. As a rule both diseases terminate in complete recovery, but in some cases there may be permanent liver injury. Also, both diseases can cause death. The IH virus kills about two out of every thousand of its victims, whereas the SH virus seems to be much more damaging, in some outbreaks killing up to half its victims.

Diagnosis is especially tricky because it is impossible to distinguish between infectious and serum hepatitis on a basis of the clinical picture alone. The patient's history is often suggestive, but in just about all instances the matter has to be worked out in the laboratory. And the most significant development here has been the finding that infectious hepatitis provokes a greater output of an antibody labeled IgM than does serum hepatitis.

The treatment of infectious and serum hepatitis is singularly nonspecific. Above all, there must be bed rest with hospitalization until the acute phase has passed and, some doctors' views to the contrary, the avoidance of excessive exercise for weeks after the patient is up and about. Prevention centers upon proper sanitation and strict control of water, food, blood, blood products, and equipment and devices employed in intravenous therapy and in making injections. For those who have been exposed to infectious hepatitis, gamma globulin has proved effective in preventing the disease when given more than one week before the expected onset of signs and symptoms. The value of gamma globulin in the prevention of serum hepatitis is still rather questionable. No vaccines are available at this time.

13

Coryza

THE THROAT IS SUDDENLY DRY AND SLIGHTLY SORE AND THE nose feels full. There is a sneeze and then another and then another and then another, and soon a watery discharge calls forth the handkerchief. Before long the eyes are bloodshot and start to water, the voice is husky, the nose congested to the point of labored breathing, and there is coughing. Meanwhile the body as a whole has given way to a concert of lethargy, malaise, and ill-described pains. Interestingly, there is no fever.

This is acute coryza (pronounced ko-ri'zah), or in common language the "common cold." Specifically, it is the customary picture of the common cold in the adult. In young children the signs and symptoms are usually much more severe and the temperature may go to 102° F. or higher.

No one really knows for sure, but the common cold is quite likely the most common malady of mankind and, further, probably causes more temporary disability than any other disease. In urban areas the general population averages about three colds per year, and in terms of economics one guess is that the common cold is responsible for a loss of some two billion dollars a year. The common cold strikes all ages and at any time of the year, though, as we all well know, the majority of the cases occur in the winter months.

The relationship between the common cold and "the virus" goes back to the year 1914, when Walther Kruse, a German bacteriologist, proved that a filterable agent was the cause of the infection. A little later Alphonse R. Dochez and his associates underscored and amplified this revelation by demonstrating filterable agents in the throat washings (from victims of the cold) that produced the disease in monkeys and human volunteers. And one is well advised to think in terms of the plural—that is, agents—because over the years one virus after another has been labeled the "cold virus"—ECHO virus 28, respiratory syncytial virus, influenza virus, parainfluenza virus, and certain reoviruses and adenoviruses. Yet, in the opinion of cold-virus pioneer Christopher H. Andrewes and others, the aforesaid viruses are not the real culprits because they generally incite *coldlike* illnesses rather than true colds. Or, put another way, when administered to volunteers these viruses do not consistently provoke the clinical picture which the doctors accept as the "common cold." This by no means implies that these viruses are any less dangerous than cold viruses; indeed, the majority are more dangerous—their coldlike ramifications merely represent either an aborted attack or a phase of a more serious repercussion. For instance, even the best diagnostician could easily label a mild case of influenza as a bad cold.

Most authorities now appear to be of the opinion that the true cold viruses are those discovered by Andrewes and his co-workers at England's famous Common Cold Research Unit in Salisbury. These agents, which Andrewes aptly refers to as rhinoviruses, can be grown easily in hu-

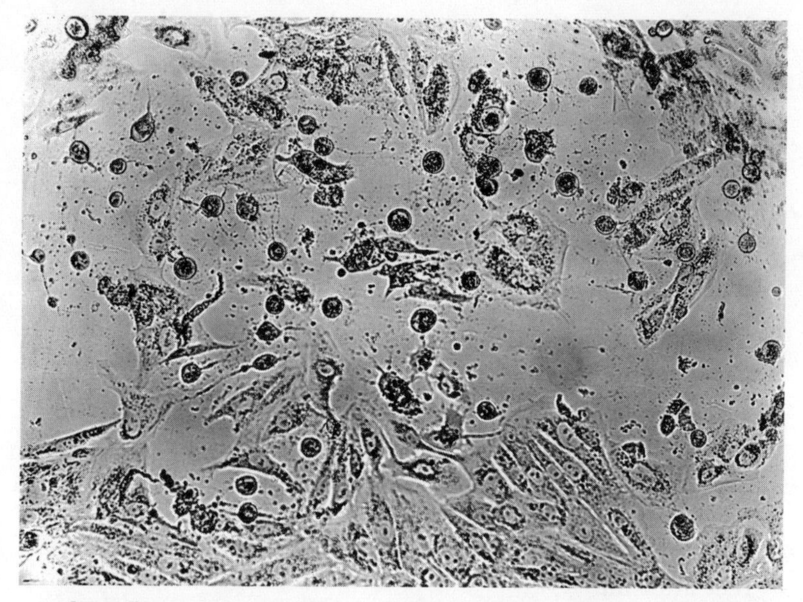

The effect of the cold virus upon living kidney cells. The affected cells are roundish as compared to the normal more or less rectangular cells along the bottom of the photo. (Courtesy Common Cold Research Unit of the National Institute of Medical Research, England.)

man embryo tissues bathed in a nutrient medium of relatively high acidity and relatively low temperature (91° F.). The acidity angle, incidentally, was the outgrowth of a happy accident, for had "medium 199" not been slightly more acid than it was supposed to be the cold virus would have failed to grow and its cytopathic effects gone unnoticed. In short, the researchers would have missed their quarry.

The rhinoviruses, which now number about two dozen or so, closely resemble the enteroviruses (which include the polio, the Coxsackie, and ECHO viruses) but differ from them in that they are rarely found in the feces. Typically, the "rhinos" multiply in the upper respiratory tract where the changes wrought in the mucosal cells are responsible for the classic signs and symptoms of the common cold. Too, they enter the blood and subsequently other tissues and thereby bring about the malaise and aches and pains.

Aside from the hassle as to what virus or viruses cause the common cold, there is also the matter of chilling, which legend ranks as a predisposing factor. The very latest researchers afford no support whatever to Grandma's admonitions about the rain, sleet, snow, and cold. But legend of this sort is a potent force and certainly there must be many doctors who go along with it. Of course, legend is not always wrong and—who knows?—maybe some virologist will come along and prove Grandma correct after all.

The diagnosis of the common cold is traditionally made by the patient or, in children, by the parent. And this can be tricky, even for the doctor. The onset of measles, German measles, or whooping cough easily passes as a cold, and so too does a mild attack of influenza or such a remote disturbance as hay fever. Since a definitive diagnosis can be made only by either the isolation of the virus or the demonstration of the appropriate antibody, we can well appreciate how many "colds" are really something else.

The views of Madison Avenue notwithstanding, the treatment of the common cold, at best, is purely and singularly symptomatic. In years past it was fashionable to give a "shot of penicillin" or some other antibiotic, but authorities are now in very good agreement that such agents should be reserved for bacterial complications. Antibiotics have no bearing on the cold virus itself, and what is more, their indiscriminate use provokes the emergence of drug-resistant bacteria. Antihistamines have also been hailed as a panacea, the contention being that if taken early enough they could abort a cold. Careful research, however, has shown this not to be true; at best this class of drugs can do little more than slow down a runny nose. To date, no one has been able to improve upon aspirin and bed rest, and the latter most particularly. And an old wives' dictum recommends that "At the first inkling of a cold hang your hat on the bedpost, drink from a bottle of good whiskey until two hats appear, and then get to bed and stay there."

The prospects for an effective cold vaccine do not appear especially bright, at least for the present time. There are far too many viruses involved and, moreover, there is evidence that these viruses are always undergoing change; for example, a vaccine prepared from virus X would not protect the body against an attack of, say virus Y, and so on. In Dr. Andrewes' words, "The net result may be that the common cold will always be with us. If so, specific vaccines cannot have a promising future."

14
Heavenly Bodies

BECAUSE OF ITS CALAMITOUS NATURE, THE ITALIANS NAMED the malady "influenza" (influence), that is, *influence of the heavenly bodies.* Though science would prove the cause to be otherwise, there is something to be said for "bodies." Since the year 1510 there have been 82 pandemics, the "Spanish flu" of 1918 being the worst. The 1918 flu appeared first in Spain as a most innocent disease and then suddenly turned overwhelmingly lethal. The notorious virus spared no area of the globe, except the mid-Atlantic island of Saint Helena, and killed an estimated 21 million people! The "Asian flu" pandemic of 1957, which appeared during the first week of April of that year and by the middle of May had spread all over the world, was much less costly but still bad enough. The death toll in the United States alone has been estimated to be in the tens of thousands.

Influenza is caused by three groups or *types* of myxoviruses—A, B, and C—which in turn are broken down into subtypes. Subtypes A (or AO), A1, and A2 were first isolated in 1933, 1947, and 1957, respectively; subtypes B (or BO), B1, B2, and B3 were first isolated in 1940, 1945, 1956, and 1962, respectively. The classification of C type virus, first isolated in 1950, is uncertain. The great bulk of cases of influenza are caused by type A viruses, and there is actually no evidence that pandemics of "B and C influenza" have ever occurred.

Above all, the various subtypes do not induce immunity to one another. For instance, when subtype A2 appeared out of the blue in 1957 no one carried antibodies against it for the simple reason that the only agent which could stimulate the plasma cells to produce such antibodies was the A2 virus itself. Thus, influenza is a formidable enemy because at any time a new type or subtype may appear against which no one is effectively protected by virtue of either a prior attack or vaccination.

What is more, a subtype itself may give way to *variant strains,* and this is exactly what happened in 1968. In July of that year an outbreak of influenza occurred in Hong Kong and spread to all age groups, resulting in the largest epidemic in that area since 1957. Viruses recovered by the National Influenza Center, University of Hong Kong, were identified as A2 variants and accordingly labeled "A2/Hong Kong/68." The new virus spread rapidly. By the end of July there were about a half million cases in Hong Kong and by the end of August there were epidemics in Singapore, Taiwan, Malaysia, South Vietnam, and the Philippines. By September, the infection had reached India, Iran, Thailand, and Australia. The virus was carried to Japan and to the West Coast by ships from Hong Kong, and by Christmas there were epidemics in just about all the states of the Northeast. The Hong Kong flu of '68 proved to be a milder affliction than the Asian flu of 1957, and the reason for this resides in the *degree of antigenic shift;* that is, whereas the antibodies against the A1 virus were about worthless against A2, the antibodies against the latter are at least partly effective against the Hong Kong variant. Put another way, the Asian flu virus and Hong Kong virus are in neighboring pews within the same church.

Some type A influenza viruses attack animals, notably birds, horses, and swine. The swine influenza virus is thought to be closely related to the subtype prevalent in 1918 and it seems likely that the pig originally contracted the infection from man. More interesting still, swine influenza virus causes the disease only in the presence of a bacterium by the name of *Hemophilus influenzae*. This was a revolutionary finding because it underscores the naïveté of necessarily assuming that microbe X, and microbe X alone, causes disease Y. A cardinal example in human medicine, the reader will perhaps recall, is diphtheria, a disease caused by a potent toxin produced by the bacterium *Corynebacterium diphtheriae*—a toxin produced *only* when the bacterium itself has been invaded by a certain virus.

Influenza occurs in sporadic interepidemic form, in periodic epidemics, and in pandemics such as those of 1918, 1957, and 1968. The epidemic periodicity, which is two to three years for virus A, apparently results from the accumulation of susceptibles and the subtraction of immune persons by reason of death; pandemics, as already noted, result from chemical changes in the virus of sufficient magnitude that there is little effective cross immunity between the pre-existing and mutated types. Pandemics occur, roughly, every decade.

The signs and symptoms of uncomplicated influenza stem from the attack of the virus on the mucous membranes which line the respiratory passageways, especially the trachea. But it must be appreciated that an attack,

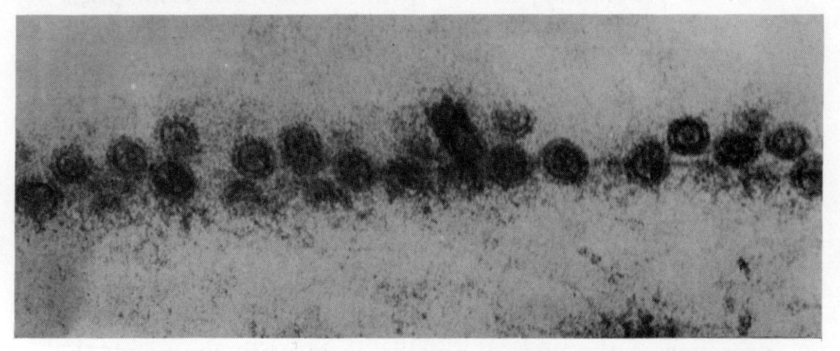

The surface of a cell infected with influenza virus and viewed at a magnification of 160,000 times. The virions are visible at the cell membrane, which passes horizontally across the field. (Courtesy, Councilman Morgan, M.D. and The Department of Microbiology, College of Physicians and Surgeons of Columbia University.)

as in all viral infections, may prove asymptomatic, and the "patient" may be quite unaware of the situation. Typically, there is dryness or soreness of the throat, a congested and runny nose, coughing, a burning pain beneath the breast bone, chills, fever, aching muscles, reddened eyes, and prostration. The temperature may rise to 106°F. and, upon occasion, there is nosebleed. The latter is of much diagnostic value. For the most part the incubation period runs for about a day and the illness itself for about ten days.

The potential deadliness of influenza arises from involvement of the lungs, either with the virus alone or with secondary bacterial invaders. Sometimes the virus attacks the lungs almost immediately upon entering the body, causing what the doctors call "primary influenza virus pneumonia." This complication, which is invariably confined to patients with existing heart trouble or lung disease or pregnancy, usually proves fatal. Bacterial pneumonias (secondary to influenza) afford a better prognosis because the causative organisms may well yield to antibiotic therapy.

Influenza can be diagnosed easily by inoculating nasal washings into the amniotic cavity of a two-week old embryonated hen's egg and then (two days later) testing for the presence of the virus by determining whether or not the amniotic fluid causes the clumping together, or agglutination, of human red cells. (Agglutination signals the presence of the virus.) Finally, to identify the particular breed of virus, hemagglutination inhibition tests are carried out using known influenza antiserums. For example, if "influenza A1 antiserum" prevents the amniotic fluid from agglutinating human red cells, it proves the presence of influenza virus A1. Another diagnostic approach is to test the patient's blood for a rising concentration of telltale influenza antibodies using known influenza viruses.

The treatment of influenza is nonspecific by virtue of the fact that the virus yields to no drug. Antibiotics, of course, may prove lifesaving in bacterial complications. In regard to prevention there has been some progress in the use of vaccines prepared from representative strains of A1, A2, and B. Briefly, the viruses are cultured in the chick embryo and then inactivated or "killed" by adding formaldehyde. Unfortunately, the immunity induced is good only for a matter of months and, too, there is no protection against the emergence of *new subtypes* and very limited protection against the emergence of *new strains.*

Older and chronically ill individuals in the population are essentially the only ones who have any risk of serious complications or fatality from influenza. Therefore, annual influenza vaccination has been recommended for them while not being recommended for the entire population. When an epidemic is forecast, however, vaccination programs might reasonably be extended beyond the high-risk groups to those providing essential community services. Otherwise, according to the Public Health Service, large-scale vaccination programs are not now warranted and should not take precedence over public health activities of already established importance. Last, but certainly not least, the vaccine should not be given to anyone who is *allergic to eggs*.

15

The Hunters

ONE OF THE GREAT MOMENTS IN MEDICINE WAS THE TRACKing down of the cause of yellow fever, an epic memorialized by Paul de Kruif's immortal *Microbe Hunters.* The hunters in this particular viral affliction became the "hunted" and many lost their lives, including the famed microbiologists Jesse W. Lazear, Hideyo Noguchi, and Adrian Stokes.

In 1881 the Cuban physician Carlos Finlay stated his belief that the mosquito carried the disease, and in 1900 the Yellow Fever Commission of the United States Army, headed by Walter Reed, proved this to be true. Also, the Commission was certain that what the mosquito really carried was a virus. In 1920 the viral nature of the disease was confirmed and years later the American physician Max Theiler developed a highly successful vaccine from the attenuated virus. For this he received the 1951 Nobel Prize in Medicine.

Yellow fever was first observed in Central America in the middle of the seventeenth century, and by the nineteenth century it had established itself throughout the Caribbean and the rain forests of Africa and South America where the disease is endemic today. In 1800 the malady hit Spain, killing 60,000 persons, and between 1851 and 1883 it killed 20,000 in Rio de Janiero and 36,000 in Havana. And in the United States there were yellow fever epidemics in New York, Philadelphia, and Baltimore, with a total in the vicinity of a half million cases throughout the nineteenth century.

In 1930 it came to light that yellow fever occurs as two epidemiological types—one called *urban* yellow fever, transmitted by the domestic mosquito *Aedes aegypti,* and the other called jungle yellow fever, transmitted by a variety of forest mosquitoes. The monkey is the chief victim of the latter type and apparently serves as an important reservoir of the yellow fever virus. In areas bordering rain forests where jungle fever is endemic, there have been epidemics of both types of yellow fever. Between 1948 and 1956 there were jungle fever epidemics in Panama and Central America and this caused many to believe that the disease would flow up through Mexico. This has not occurred yet to any significant degree but it certainly remains a distinct possibility in the face of laxity in preventive matters.

The yellow fever virus belongs to the arbovirus family, a group characterized by their affinity for the salivary glands and bodies of mosquitoes and ticks. Kipling's reminder about the female being more deadly than the male is exquisitely apropos here because the female mosquito must drink blood to nourish her eggs. Specifically, when the female Anopheles mosquito bites a victim of yellow fever, she picks up the virus and passes it on to a susceptible person in a subsequent bite. Since the virus is present in the blood of man only during the initial three or four days of the infection, the mosquito must attack during this period to become infectious. Further, a week or so

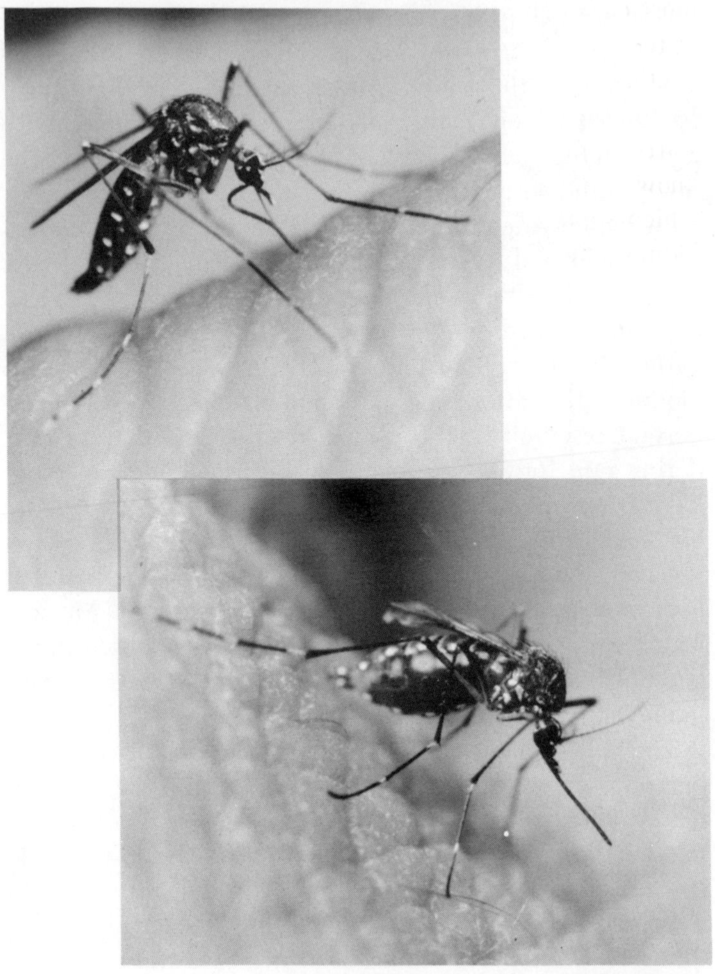

Feeding on human blood from index finger, a female yellow-fever mosquito appears above. The needle-like part of the proboscis is piercing skin. Below the mosquito is seen with abdomen swollen from the blood meal. Only the female insect drinks blood. (Courtesy Thomas Eisner of Cornell University.)

must elapse in order for the virus to become established in the mosquito's salivary glands.

Once the virus has been injected into the body, either by the Anopheles mosquito or one of the other species, it starts to multiply and in five days or so makes it presence known through an array of signs and symptoms, all of which apparently arise from the attack of the virus on the kidneys, heart, spleen, and, most especially, the liver.

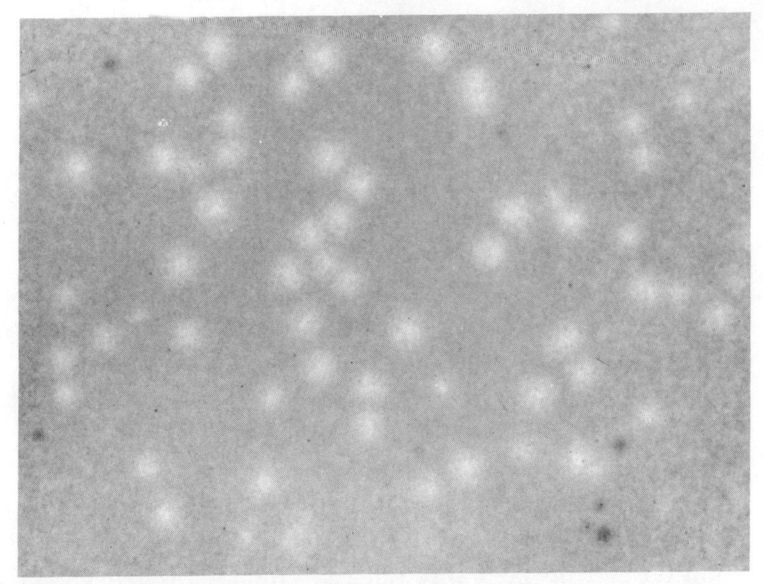

Tissue culture showing plaques (areas of destruction) caused by the yellow fever virus. (Courtesy National Medical Audio-visual Center, Atlanta, Georgia.)

Degenerative changes are noted in these organs and the destroyed cells are so typical in their appearance that they were given the special name of Councilman bodies (after their discoverer William T. Councilman).

The so-called first stage of yellow fever is marked by severe headache, flushed face, reddened eyes, pain in the back and legs, prostration, nausea, vomiting, and a fever that may shoot up to 105°F. or more. Protein appears in the urine and sometimes there is a sharp drop in the white-cell count. On about the second day these ominous rumblings clear up only to be followed on the fourth or fifth day by jaundice, bleeding gums, nosebleed, and hemorrhaging beneath the skin and along the gastrointestinal tract. In fatal cases the characteristic features are dark vomit and tarry stools, cessation of urination, and hiccups. Terminally, the patient passes into coma and usually dies in a week's time. Sometimes death is heralded by wild delirium.

The foregoing is yellow fever in all its classic horror; most fortunately, it is not the picture noted in the majority of cases, popular beliefs notwithstanding. Indeed, not uncommonly, the only symptom is headache. In some epidemics fatality rates have gone over fifty percent, but the overall average is put by most authorities below ten percent. Still, this is high and it is also disconcerting to learn that what appears to be a mild attack may suddenly change for the worse. For the most part, though, if the early signs and symptoms are mild the outcome is good and a rapid recovery can be expected. Severe cases can and do end up favorably but the prognosis is not good, particularly when underscored by black vomit, tarry stools, and noisy hiccups.

A diagnosis based upon clinical findings is not exactly easy, although a concert of high fever, black vomit, protein in the urine, and jaundice are highly suggestive. Infectious hepatitis, serum hepatitis, malaria, and even carbon tetrachloride poisoning can mimic yellow fever. A definitive laboratory diagnosis centers upon the isolation of the virus in mice and monkeys and the demonstration of telltale antibodies in blood specimens.

In the way of treatment there is little the doctor can do but put the patient to bed and administer analgesics and intravenous fluids. Preventionwise, though, much can be done. First and foremost, an effective vaccine is available in the form of an attenuated virus (Theiler's famous 17D strain). Vaccination—a must for persons living or entering endemic areas—provokes an immunity in a week's time that lasts for about six years. Mosquito control is obviously a key factor in prevention and so, too, is keeping the patient under a bed net or in a mosquito proof room during the first week of the illness. And last but by no means least Nature has had mercy upon us by providing a lifelong immunity in the wake of an attack of the disease.

16

St. Louis Blues

IN 1930 A STRANGE MALADY STRUCK GREAT NUMBERS OF horses and mules throughout California's San Joaquin Valley. The animals became drowsy and feverish and some went into convulsions. Autopsies disclosed inflammation (*itis*) of the brain (*encephalo*) and spinal cord (*myel*), and the disease was accordingly labeled equine *encephalomyelitis.* In 1933 a similar but more deadly disease hit Virginia and New Jersey. Horses and mules, once again, were the victims. K. F. Meyer and his co-workers had isolated a particular virus from the brains of the afflicted animals in the California epidemic and now Charles Ten Broeck and his associates isolated a virus in this epidemic of 1933. In 1938 an epidemic of encephalomyelitis in horses and mules occurred in Venezuela and, once again, a virus was isolated. The agents involved in the three outbreaks proved to be distinct viruses and the diseases therefore were given the somewhat unwieldly appellations of *western* equine encephalomyelitis, *eastern* equine encephalomyelitis, and *Venezuelan* equine encephalomyelitis.

Eastern equine encephalomyelitis—EEE—struck again in 1938 in Massachusetts, this time attacking man as well as animals. Of the 40 human cases reported, the bulk of

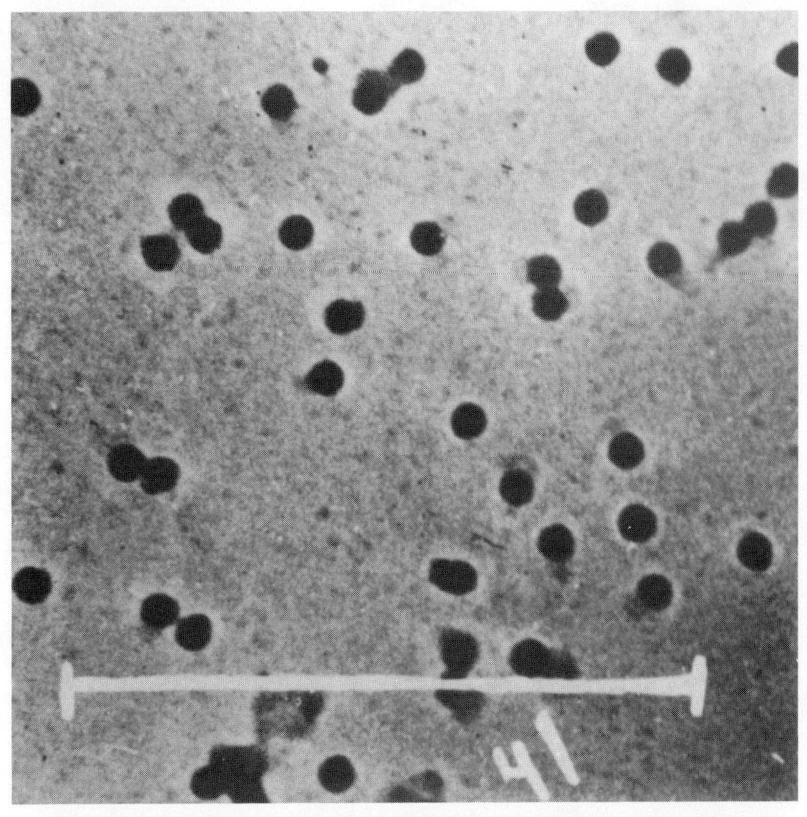

The virus of Eastern equine encephalomyelitis. The white line represents 1000 millimicrons, which means that each virion is about 50 millimicrons in diameter. (Courtesy of the National Medical Audiovisual Center, Atlanta, Georgia.)

which were young children, 26 were fatal. In 1959 there was an outbreak of the eastern variety in New Jersey, killing 21 out of 32 confirmed cases in humans, and again in 1968 killing 6 out of 12. The virus of western equine encephalomyelitis—WEE—hit the human population of the North Central states in 1941, causing some 200 deaths among more than 3,000 cases. Subsequent attacks of the western type have resulted in similar mortality rates (about 7 percent), demonstrating quite dramatically that in man EEE is more of a killer than WEE. Interestingly, in the horse the two infections are substantially the same. Venezuelan equine encephalomyelitis—VEE—also occurs in humans, but to date has confined itself to Central America and South America.

The causative agents of the three kinds of equine encephalomyelitis are mosquito-borne arboviruses. The natural reservoir of the infection is in wild birds; the viruses have been isolated from well over forty species. Too, there is evidence of infection in snakes and domestic birds and mammals. Despite its name, therefore, equine encephalomyelitis is primarily a disease of birds, with man, the horse, and the mule playing an "incidental" role during periods of high prevalence. Dozens of species of mosquitoes serve to transmit the viruses from infected birds to man and animals.

Once the WEE virus enters the blood the signs and symptoms of infection appear in about one to three weeks. Typically, these include headache, drowsiness, fever, mental confusion, and amnesia. The clinical picture is not uncommonly without symptoms and then at the other extreme there may be convulsions and paralysis. But in any event the acute phase lasts for about a week, with the great majority of cases resulting in uneventful and complete recovery. The EEE virus produces a similar type of infection save for the signal fact, as already noted, that it is roughly ten times more deadly. Put another way, the diseases are alike qualitatively but not quantitatively.

Venezuelan equine encephalomyelitis in man differs greatly from the eastern and western types. After a relatively short incubation period of three days or so there is a phenomenally abrupt onset of illness that generally mimics a bad cold or the flu. There are headache, fever, shaking chills, muscle pains, prostration, and all the rest. Happily, VEE is self-limited and the mortality rate is thought to be well under one percent.

In 1933, the same year of the EEE outbreak in Virginia and New Jersey, a weird sort of disease hit St. Louis and surrounding areas in epidemic proportions. Of 1,130 reported cases there were over 200 deaths. Sometimes the illness involved no more than a headache and fever and then again there was a concert of ominous repercussions—severe abdominal pain, sore throat, reddened eyes, difficulty in walking, mental confusion, and, not uncommonly, paralysis. Worse yet, some cases in infants resulted in permanent brain damage; recovery in older children and adults was usually complete. As one might suspect most of the deaths occurred in the very young and the very old. Autopsy findings revealed extensive inflammation of the brain and the new disease was immediately labeled "St. Louis encephalitis." Soon public health workers isolated the etiologic agent, which has now been demonstrated to be another arbovirus.

Over the years the St. Louis encephalitis—or SLE—virus has been recovered from several species of domestic and wild birds and there is evidence of infection in mammals. Thus, as in the case of equine encephalomyelitis the bird serves as the reservoir of infection and, once again, the mosquito serves as the vector. Interestingly, there is no evidence that the SLE virus attacks the mule or horse. Since the 1933 outbreak there have been sporadic flareups and small to moderate epidemics in Texas, California, and the lower Ohio River Valley.

Though WEE, EEE, and SLE are the major forms of viral encephalitis in the United States, the rest of the world is plagued by other arbovirus brain involvements—namely, Japanese B encephalitis, Murray Valley encephalitis, and Russian spring-summer encephalitis. There is no specific treatment for any kind of encephalitis and, for the most part, vaccination is either ineffective or impracticable. A killed vaccine does a good job in the prevention of WEE and EEE in horses and mules, and on a limited basis it has been used in persons under unusual risk, such as veterinarians and laboratory workers. For the present, anyway, the best approach is to avoid and destroy mosquitoes.

17

Campobello

ON A WARM SUNNY DAY IN AUGUST IN THE YEAR 1912, A young man slipped and fell overboard while sailing. A slight chill notwithstanding, the following day he went for a swim in the lake, jogged a mile and a half, and finally took a dip in the icy waters of the Bay of Fundy. That afternoon at his summer retreat on Campobello Island he sat in a wet bathing suit for an hour or so reading his mail. Toward evening there was another chill and the next day he had pain in his back and legs and a high fever. A local doctor said it was a cold, but by then the patient could neither walk nor move his legs. An "expert diagnostician" was called in and it was his opinion that there was an injury of the lower spinal cord. Two weeks later another diagnostician made another diagnosis—poliomyelitis or, as they used to say, infantile paralysis.

Franklin Delano Roosevelt was in agony. At one time his arms and back were paralyzed and for weeks he could neither move his bowels nor pass his urine. He suffered terrible mental depression and the doctor in charge of the case remarked that "it will take all the skill we can muster to lead him successfully to a recognition of what he really faces without crushing him."* It took FDR years to realize that he would never walk again, but thanks to an indomitable spirit, an understanding and inspirational wife, and the salubrious waters of Warm Springs, Georgia, a lowly virus was thwarted in its attempt to alter the course of history.

Though poliomyelitis is unquestionably ancient, the first description of the disease did not appear until the year 1789, when Dr. Michael Underwood of England took pen in hand and rendered what in all likelihood was a true case of the dread disease. In 1840, Jacob von Heine of Germany recognized polio as a clinical entity, and in 1891 Karl Oskar Medin of Sweden described what is generally regarded as the very first epidemic. The first major epidemic in the United States occurred in 1916, and from about this time on there has been a shift from the endemic form (characterized by sporadic cases) to epidemic in all highly developed countries.

Endemic polio continues to persist in primitive populations, particularly in tropical areas. Too, there has been a shift from zero to four in the peak-age incidence to five to nine and an increasing number of cases in older children and adults. Apparently, these shifts arise from the postponement of first exposure to the virus under improved socioeconomic conditions. For example, whereas most persons in areas of poor sanitation develop antibodies in early childhood, peak population immunity is not reached until fifteen or thereabouts in areas of good sanitation. Again, urban dwellers are more likely to become immune earlier than those who live in rural areas because crowding obviously enhances the spread of the virus.

---

* Burns, J. M., *Roosevelt: The Lion and the Fox.* New York: Harcourt, Brace and World, 1956

Other factors having a bearing on polio include pregnancy, menstruation, absence of tonsils and adenoids, chilling, and physical exertion—all of which are believed to heighten susceptibility to the disease. Sex, too, plays a role. Whereas among children males are more affected than females, just the opposite is true in adults. The role of the fly is still not settled. Flies trapped in epidemic areas may carry the virus and experiments have shown that they can transmit the virus to monkeys by contaminating food, but fly control programs have yielded equivocal results. In regard to the season of the year, polio is most common from July through September in the temperate zone, although it may appear as early as April and as late as December.

Man is the sole reservoir of the polio virus and its sole natural victim. The chimpanzee, however, is quite susceptible and has in fact been infected accidentally in the laboratory. The disease both in this animal and the monkey closely resembles that in man. Although for years it was thought that only primates could be experimentally infected, in 1939 the virus was "adapted" to mice, rats, and hamsters.

We speak of the "polio virus," but actually there are countless viruses involved, the first of which was isolated by Karl Landsteiner and Erwin Popper in 1908 by inoculating monkeys with nervous tissue from a fatal case in a human being. These viruses belong to the family of enteroviruses because all are transient inhabitants of the intestinal tract. Moreover, and most fortunately, polio viruses fall into three so-called types—Type I, Type II, and Type III (known also as Brunhilde, Lansing, and Leon, respectively.) Brunhilde was the name of the chimpanzee from which the "original" virus was isolated; Lansing is for

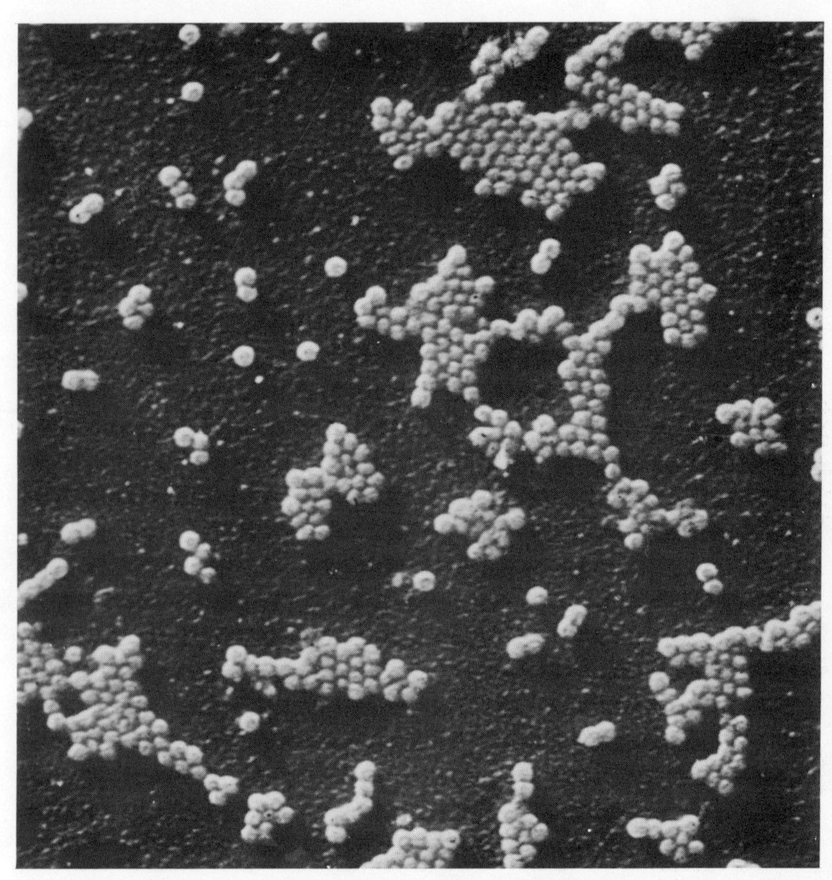

The poliovirus (type 2) magnified 79,000 times. (Courtesy Park, Davis and Company.)

Lansing, Michigan, where the patient lived from whom this type was first isolated; and Leon is named in memory of the Los Angeles boy from whom the specimen was obtained. These viruses show up in the nose, throat, blood, and stool specimens of acute and convalescent cases, in the spinal cord and brain of fatal cases, and in sewage and flies during epidemics. The spread and mode of infection is thought to be almost entirely fecal-oral. The virus is recoverable from patients' feces for several weeks and a small quantity of stool has been demonstrated to contain thousands of disease-producing doses of the virus. In sewage or water the virus remains viable for as long as four months.

The isolation and cultivation of polio viruses by inoculating monkeys is exceedingly slow and extremely expensive and fraught with obvious technicalities. Indeed, it was not until Enders, Weller, and Robbins adapted the virus to the tissue culture that poliomyelitis began to yield its secrets. Tissue cultures, as we saw earlier, afford an easy means of isolating and identifying a virus and an economical way of growing a virus in bulk for vaccine production.

The polio virus enters the mouth, multiplies in the throat and intestinal tract and from these sites may pass into the blood and nerve tissue. The exact path is not known with absolute certainty, but most authorities believe that once in the blood the virus makes its way into the brain and spinal cord by direct passage from the capillaries. The neuron, or nerve cell, provides an immediate home for the polio virus and in doing so often sets the stage for its own destruction. This is particularly true of the neurons of the front portion of the gray matter of the spinal cord, a fact which accounts for polio's full medical name—*acute anterior poliomyelitis* (anterior—front; polio—gray; -myelitis—inflammation of the spinal cord). The brain and spinal cord are congested and slightly swollen and tiny hemorrhages may be seen. Additionally, in the more chronic stages of the disease there are degenerative changes in the nerves and muscles.

The phenomenal fact about polio, and the one not generally known to most people, is that only about one percent of the cases of the disease are recognized for what they are; 99 percent appear as inapparent infections, mild nonspecific diseases, or disorders producing signs of meningeal irritation without paralysis. (The meninges are the three membranes enveloping the brain and spinal cord.)

Most authorities recognize four categories of polio: inapparent infection, "minor illness," nonparalytic, and paralytic. In inapparent polio the person harbors the virus and yet is unaware of any illness. Perhaps 95 percent of the "cases" are of such form. Time and time again the virus has been recovered from the feces and throat of apparently healthy members of families in which a case of polio has occurred. In the minor illness the clinical picture is one of a nonspecific character involving either the upper respiratory or gastrointestinal tract. Just how many cases of the grippe are in reality a case of polio minor illness is both an interesting and moot question.

Once the virus invades the nervous system the minor illness typically gives way to meningeal irritation characterized by headache, stiff neck, and stiff back. The infection may resolve itself at this stage, and if it does, the label "nonparalytic polio" is applied. If it does not resolve itself, paralysis ensues and the label is "paralytic polio." In *spinal* paralytic polio the neurons of the spinal cord have been attacked and the muscles affected are those inner-

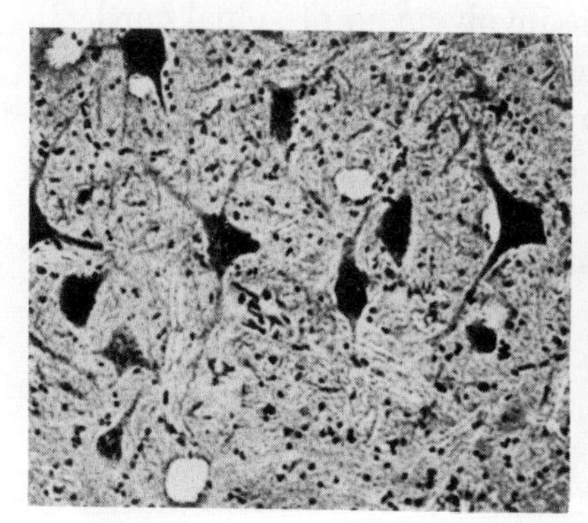
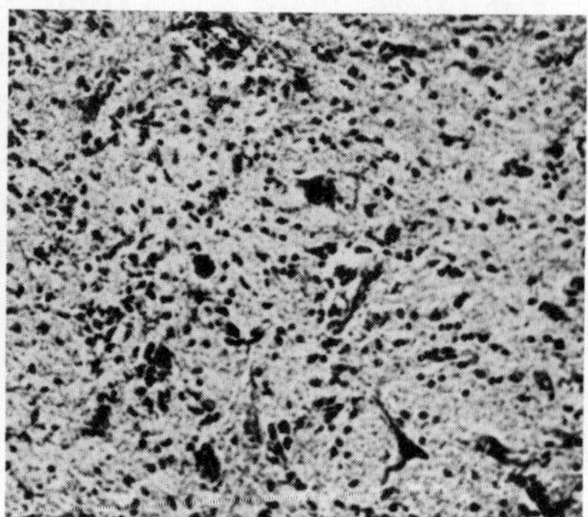

The normal spidery nerve cells (left) of monkey spinal cord are all but obliterated (right) by the poliovirus. (From Burrows, W.: *Textbook of Microbiology,* ed. 17, Philadelphia, 1959, W. B. Saunders Company.)

vated by the nerve fibers supplied by the deranged neurons. Usually, cramping pains are noted first, then muscle weakness, and finally paralysis. In children under five years of age the most common feature is a slight paralysis of one leg; in children between five and fifteen years weakness of one arm or paralysis of both legs (paraplegia) is typical; and in adults paralysis of all four limbs (quadriplegia) is seen most often. Paralysis of the muscles of respiration is more common in adults than in children and so too is disfunction of the urinary bladder.

In *bulbar* paralytic polio the virus attacks the neurons of the brain stem and the clinical picture depends upon the particular areas involved. Restlessness, anxiety, falling temperature, poor balance, deafness, difficulty in swallowing, inability to speak, low blood pressure, shock, and respiratory failure are all possibilities. The latter is invariably the cause of death. And here it is of practical interest to note that whereas the respiratory failure in spinal paralytic polio arises from the involvement of spinal cord neurons whence arise the nerves supplying the muscles of respiration, in bulbar polio "only" the *respiratory center* in the brain is knocked out. This is the center that initiates respiration. Thus, in bulbar respiratory failure *artificial* electrical stimulation of the nerves supplying the respiratory muscles will cause them to contract; in spinal respiratory failure these nerves are damaged and therefore unresponsive.

Actually, there is little stopping the virus from attacking both the brain and the spinal cord, and *bulbospinal* poliomyelitis does occur. Another possibility is polioencephalitis, which results from an attack of the virus upon the higher brain centers. This may occur as an isolated situation or along with spinal, bulbar, or bulbospinal disease. Very rarely, epidemics occur in which most of the cases are of this kind. Characteristic features are quivering, trembling, twitching, jerking of facial muscles and extremities, flushing of the face, anxiety, a feeling of impending doom, and severe insomnia. In fatal cases severe confusion progresses to lethargy, coma, and ultimately death.

An unequivocal diagnosis of poliomyelitis can be made only on the basis of a *rising* concentration of antibodies in the blood as disclosed by specimens taken in the acute and convalescent stages. Polio antibodies are of the neutralizing kind and prevent the virus from destroying cells in tissue culture; that is, they inhibit the virus's cytopathic effects. By way of example, if a blood specimen inhibits the effects of Type I polio virus, this shows the presence of Type I antibodies, meaning that the body has encountered Type I virus. As indicated, however, unless the concentration of these antibodies is on the rise (as disclosed by more antibodies in the "convalescent sample" than in the "acute sample") it means that the Type I infection occurred some time in the past. Thus, a rising concentration of Type I antibody spells out a diagnosis of poliomyelitis caused by Type I polio virus. In the same way a rising concentration of Type II and Type III antibody concentration affords a positive diagnosis of Type II and Type III poliomyelitis, respectively.

In making a diagnosis on a purely clinical basis the doctor is obviously on shaky ground. In the inapparent infection there are no signs and symptoms and in the minor illness the signs and symptoms are so nonspecific that the underlying cause is completely overlooked except in an epidemic. As for a diagnosis of nonparalytic polio the best the doctor can do at the bedside is to prove or disprove the presence of meningeal irritation, a situation, as already noted, that can be set off by a great many microbes other than the polio virus, not to mention a variety of chemicals and drugs. Still, the presence of meningeal irritation during a polio epidemic is highly suggestive and routinely tested for in suspected cases. The standard clinical tests for meningeal irritation are Kernig's sign and Brudzinski's sign. In the former the leg cannot be completely extended when the thigh is flexed against the abdomen (in the lying position), and in the latter flexure movements of the ankle, knee and hip are produced when the neck of the patient is bent.

The clinical diagnosis of paralytic polio is equally tricky because of the many mimickers. Brain tumors, hemorrhage, blood clots, acute multiple sclerosis, acute rheumatic fever, trichinosis, and syphilitic meningitis are a few possibilities, not to mention a variety of viral involvements of the central nervous system. The measles virus, for instance, may attack the brain in such a way that the signs and symptoms may compose a "picture of polio." But again, paralysis—along with signs of meningeal irritation—is highly suggestive of acute poliomyelitis during an epidemic, and the seasoned diagnostician is usually right when he makes such a diagnosis.

The treatment of polio ranges from doing nothing at all to elaborate medical and nursing measures, depending upon the nature and severity of the attack. The most pressing cases are those of respiratory failure and where the patient cannot swallow. For the respiratory failure of spinal origin a tank respirator, or "iron lung," may be needed to effect breathing, whereas in bulbar respiratory failure the electrophrenic respirator is used. In the latter device an electrode is placed on the neck and a tiny stimulus applied at a rate of about twelve to fifteen times per minute, each shock causing the phrenic nerve "to fire" and contract the diaphragm. For example, a respirator set at 15 would effect 15 respirations per minute, just about the normal rate. Other therapeutic considerations include the use of antibiotics to control bacterial infections which may arise (usually of the lungs and urinary tract), physiotherapy to "maintain" the weak muscles until nerve function returns, and the control of emotional disorders which often accompany a severe attack. Important, too, is the role of occupational rehabilitation in the instance of those patients who are unable to resume their work.

The outlook in a case of polio is relatively hopeful. Patients with minor illness and nonparalytic polio recover completely, and so do many with paralysis. The return of some degree of muscle function is almost always a rule and very few remain totally paralyzed. Paradoxically, statistics show that the more severe the infection is in the acute stage, the better the chance for complete recovery. For example, difficulty in swallowing and respiratory failure generally disappear completely. As one would suspect, bulbospinal polio affords the poorest prognosis, with a fatality rate running as high as 75 percent. For all forms of paralytic polio the fatality rate rises with age, averaging about 3 percent in children in contrast to 20 percent in adults. Overall the polio virus kills about five out of every hundred of its victims. Pathology tells us that recovery is complete when no neurons are destroyed; functional return, or degree of recovery, depends upon the *number* of neurons destroyed.

Perhaps the most dramatic development in medical virology of the present century resides in the polio vaccine. The inspiration for and the operation of the vaccine relates to the fact that an attack by a polio virus confers a lifelong immunity against a reinvasion by the *same type* of virus. And it makes no difference whether the attack is mild, severe or even inapparent. Thus, an "intestinal upset" due to Type I polio virus is just as effective in provoking antibodies against reinvasion as a bout of paralysis. This explains why the great bulk of the adult population carries protective antibodies even though "they have never had polio." In short, by the time most people reach fifty years of age they have encountered Types I, II, and III polio viruses and are protected against polio in any form.

In 1954, in the most stupendous human experiment ever conducted, a vaccine perfected by Jonas E. Salk and his associates at the University of Pittsburgh was proved "completely effective" in provoking protective antibodies without the risk of either killing or crippling. Fittingly, this pronouncement was made on April 12, 1955, the anniversary of the death of Franklin Delano Roosevelt. The figures speak for themselves. Whereas in the prevaccine year of 1952 there were 57,000 cases of polio, in 1963 there were only 431 cases!

The Salk vaccine is prepared from formaldehyde-*killed* polio viruses of the three types. It does a good job against paralytic polio, but follow-up studies have demonstrated the need of repeated boosters to maintain a protective level of antibody. For these reasons medical authorities believe that *live* vaccines are of greater promise in the prevention of the disease. The pioneer Sabin oral vaccine (prepared from carefully selected, tamed viruses) produces what amounts to be a "natural" inapparent infection resulting in a solid immunity against both paralytic and nonparalytic polio. Some concern has arisen because the tamed virus on a few occasions has apparently caused a paralytic illness, but this is generally not regarded as significant when one considers that hundreds of millions have been vaccinated without incident.

18

Ammon's Horn

"SURE ENOUGH, LESS THAN TWO WEEKS LATER THE GOOD creature began to howl mournful cries and tear up his bed and gnaw his cage—and in a few days this brute died so that thousands of mankind might live. . . ."*

The "good creature" . . . "this brute" . . . had been injected—right in its brain—with a bit of the ground-up brain of another dog just dead of rabies. And so in the year 1882 Louis Pasteur and his associates, Pierre Roux and Charles E. Chamberland, demonstrated that the unseen deadly virus settles in the nervous system. What is more, they saw in this an easy means for growing the virus in bulk amounts for vaccine purposes. The trick, of course, was to tame the virus, and this the three workers did by removing a piece of spinal cord from a rabbit dead of rabies and drying it out for two weeks. The powdered bit of shriveled cord failed to incite the disease in dogs when injected into their brains and thus Pasteur reasoned that the strongest immunity could be provoked by giving a *series* of injections, starting with the highly weakened fourteen-day-dried virus and ending with the highly active

* deKruif, P., *Microbe Hunters,* New York: Harcourt, Brace and World, 1953.

one-day-old virus. Two dogs were given the lengthy series —without ill effects—and then one day weeks and weeks later both were inoculated with a heavy dose of lethal virus. The animals never showed the slightest sign of rabies!

Still, these were dogs, not humans, meaning that the ring had to be closed in the body of man. And this was an awesome prospect because there had never been a proved instance of recovery in man from rabies. Pasteur was all set to play the guinea pig himself, but as fate would have it on the night of July 6, 1885, an hysterical mother came into his laboratory with the plea, "Save my little boy!" Two days earlier nine-year-old Joseph Meister had been bitten severely by a "mad dog," and if the dog had indeed delivered rabid saliva beneath the skin Joseph was as good as dead. On the other hand, who was to say with certainty just how safe were those rabid rabbit spinal cords? Be that as it may, Joseph Meister received the fourteen injections and lived a long and happy life.

This dramatic discovery not only afforded a clear-cut control over the most deadly infection known to man but also heralded *attenuation*—or "taming of a virus"—as the most potent means of provoking artificial immunity.

Rabies is one of the oldest known maladies. Aristotle recognized it as a disease of animals and the celebrated Roman physician Celsus first described its transmissibility to man. Matthieu Orfila in 1809 demonstrated that the saliva of rabid animals was the true source of infection; Victor Galtier in 1879 described the susceptibility of rabbits to rabies and use of the animal for diagnostic purposes; and as we have seen, Pasteur closed the ring between 1881 and 1885.

All warm-blooded animals—and man—are susceptible

to rabies, and outbreaks of the disease occur in any climate during any season of the year. So-called "natural," or sylvatic, rabies refers to the disease in wildlife and the so-called "urban" type refers to the disease in domestic animals, most particularly the dog. Wolves, foxes, coyotes, skunks, and bats represent the natural reservoir of infection and it is from these animals that the disease is passed onto the dog. The bulk of human cases stem from dog bites, and with the elimination of canine rabies there is no evidence that the disease persists in urban areas. Relative to rabies, then, the dog is far from being man's best friend.

And the bat may well be the worst enemy of all, for there is growing evidence that it constitutes the major reservoir of the virus in the Western Hemisphere. Above all, rabies is uniformly fatal to all animals except bats, many of which survive the disease to become healthy carriers. Bat rabies was first recognized in the vampire bat in Brazil in 1908, and since 1953 the disease has turned up in a number of nonblood-eating species in various parts of the United States. Vampire bats bite at night to transmit the infection to cattle and other domestic animals and to man; within an affected area as many as half the cows may be involved. Between 1929 and 1931 a thousand cows died in Trinidad, and in 1929 there were thirteen human deaths also ascribed to the vampire bat. A 1957 survey in Florida underscored the bat problem in this country; and in one episode in which twelve bats had bitten animals, four were found to be infected.

The virus of rabies has now been isolated, measured, tested, and categorized as a myxovirus, the family that also includes the viruses of influenza, measles, and canine distemper. The chief feature of the family is the ability

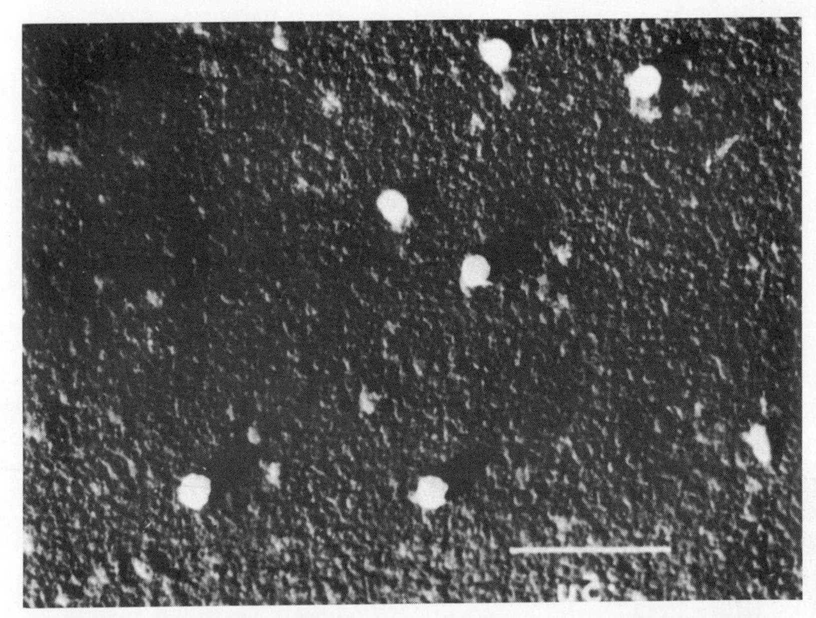

Rabies virus. (Courtesy National Medical Audiovisual Center, Atlanta, Georgia.)

of member viruses to clump together, or agglutinate, red blood cells. Too, the family is characterized by its content of RNA and sensitivity to ether. The rabies virus cannot pass through intact skin but there is good evidence that it can pass through the intact mucous membranes of the nose and throat. Commonly, the virus is present in the saliva of rabid animals, especially during the incubation period; it is usually transmitted via bites and licks. Recently, researchers have turned up evidence of airborne infections in caves inhabited by rabid bats.

Contrary to what many believe, a bite inflicted by a rabid animal transmits the virus in only about half the cases. One interesting study involving rabid wolves—generally considered the most dangerous purveyors of the disease—showed that out of 32 persons bitten by the same rabid animal (and who did not receive the vaccine), only 15 contracted the disease and, subsequently, died. To a certain extent the attack rate depends upon the location and severity of the wound and whether or not the bite was inflicted through heavy clothing. Bites on the head and neck are without question the most dangerous of all, for in these areas the virus is afforded easy access to the brain and spinal cord via a galaxy of nerves. The importance of the neural pathway is underscored by the finding that cutting an infected nerve blocks the spread of the virus.

The incubation period of rabies varies from a week or so to *a year or more,* the length apparently being related to the severity of the injury and the amount of virus introduced at the time of exposure. The onset of the disease generally includes loss of appetite, malaise, headache, fever, sore throat, and a tingling and burning pain around the site of the wound. The skin becomes ultrasensitive to temperature changes and drafts, and quite typically there is acute sensitiveness to light and sound. In the majority of cases as the disease progresses there are muscle spasms of the mouth and throat on drinking, and eventually the mere sight of water or any fluid will convulse the throat and voice box—hence, the term *hydrophobia* ("fear of water"). The victim becomes irrational and often maniacal, the pathological event that prompted someone to alter the Latin *rabere* ("to rage") to rabies. This so-called "excitation phase" reaches its acme in convulsions so terrible that the body may be thrown into *opisthotonus,* a form of tetanic spasm in which the head and heels are bent backward and the body forward. Death may arrive in this grotesque posture or, as in many cases, not until the convulsions give way to a generalized paralysis. The clinical picture obviously varies somewhat from victim to victim except in regard to one point—the disease is always fatal.

At autopsy the brain is congested, puffy, and crumbly and the convolutions are broad and flattened. The salivary glands are usually soft and swollen. In the majority of cases the microscope discloses inclusion bodies within the nerve cells, especially those cells in that portion of the brain called the hippocampus, or Ammon's horn. These inclusion bodies, generally called Negri bodies in honor of their discoverer Adelchi Negri, are emblematic for rabies and in their absence the aforesaid brain lesions could pass for any number of other viral infections of the nervous system.

Because no one has ever recovered from rabies there is *no* treatment in the strict sense of the word—there is only "protective treatment." And the sole reason this protective treatment exists is because of the phenomenally long incubation period—an incubation period long enough in most cases to allow the body's plasma cells to produce sufficient antibody to neutralize the virus before it reaches the brain and spinal cord. In a very real sense the rabies victim must beat the clock or die.

And nowhere in the practice of medicine are the rules of "what to do" spelled out in such a definitive way as in the handling of an animal bite. First, and perhaps in many instances foremost, all bite wounds, including scratches and other skin abrasions exposed to animal saliva, must be cleansed immediately with *plenty of soap*. Tests have demonstrated conclusively the superiority of soap solutions over just about anything else and it behooves everybody to digest this fact and never forget it. This is especially true for those living in remote areas.

What to do next falls in the province of the doctor and he in turn follows the basic guide promulgated in 1957 by the Expert Committee on Rabies of the World Health Organization. This guide is predicated upon the condition of the biting animal at the time of exposure and during the ensuing ten days. The specifics are most critical and, to say the least, prove interesting reading.

CATEGORY ONE

This includes all cases of *indirect* contact and *no* visible lesions. If at the time of exposure the animal was rabid no treatment is considered necessary except in young children and in patients for whom a reliable history cannot be obtained, in which case vaccination must be started immediately. Vaccination follows the Pasteur regimen of an injection per day for fourteen days.

CATEGORY TWO

This includes *all licks* and four possibilities. A lick on intact skin by a rabid animal calls for no vaccination except in young children and in patients for whom reliable history cannot be obtained, in which case vaccination must be started immediately. A lick on abraded skin and abraded or unabraded mucous membranes calls for no vaccination if the animal is healthy at the time of exposure and is healthy during an observation period of ten days. If the animal shows clinical signs of rabies or is proved rabid, vaccination is started immediately at the first sign of the disease in the animal. If the animal shows signs suggestive of rabies at the time of exposure vaccination is started immediately and stopped if the animal is normal on the fifth day after exposure. If the lick was delivered by an animal that escaped, was rabid, killed or unknown, vaccination is started immediately.

CATEGORY THREE

This includes *all bites* and eight possibilities. A *minor* bite calls for no vaccination if the animal is healthy at the time of exposure and during an observation period of ten days. If the animal shows clinical signs of rabies or proved rabid during the observation period vaccination is started immediately at the first sign of the disease in the animal. If the animal shows signs suggestive of rabies, vaccination is started immediately and stopped if the animal is normal on the fifth day. If the animal is rabid at the time of exposure or escaped, or killed or unknown and if the bite is by a wolf, jackal, fox or other wild animal vaccination is started immediately.

Severe bites—multiple or a bite on the face, head or neck—by an animal healthy at time of exposure and healthy during the ten-day observation period calls for an immediate dose of hyperimmune serum (that is, blood serum containing antibody against rabies) but no vaccine. If the animal was healthy at the time of the exposure but shows clinical signs of rabies or proved rabid during the observation period hyperimmune serum is given immediately and vaccination started at the first sign of the disease in the animal. If the animal shows signs suggestive of rabies at the time of the exposure hyperimmune serum is given followed by vaccination—and the latter may be stopped if the animal is normal on the fifth day after exposure. Finally, a severe bite inflicted by a rabid, escaped, killed, unknown, or wild animal calls for hyperimmune serum immediately followed by vaccination, that is, the "full treatment."

The cardinal points for the man on the street, then, are these:

1. Wash the wound with plenty of soap and water.
2. Notify the police or other authorities of the incident and the whereabouts of the animal (which is turned over to a veterinarian for the observation period).
3. Get to a doctor at once.

The uneventful vaccination of Joseph Meister notwithstanding, Pasteur's rabid-rabbit spinal cord preparation was not without danger and not a few patients were killed by it. A number of modifications were made of the original but all those derived from nerve tissue have on occasion provoked an allergic involvement of the brain and spinal cord resulting in paralysis. Indeed, this is the reason why it is most important to avoid use of the vaccine if at all possible, and why the recommendations set forth by the Expert Committee on Rabies are so critical in their application.

The safest preparation available at present is a killed vaccine of duck-embryo origin; that is, the embryo is infected with the rabies virus and then dried, powdered, and chemically treated to deactivate the virus. This vaccine contains little or none of the "paralytic factor" said to be present in the vaccine made from brain tissue. Though some authorities feel that this vaccine still has to prove itself to be as good as the older "nerve preparations" there is both laboratory and bedside evidence of its effectiveness in stimulating antibodies speedily enough to protect an exposed individual.

And another rather recent development is the hyperimmune serum referred to earlier. Prepared from the blood of vaccinated individuals, a shot of the stuff provides an immediate supply of antibodies, apparently enough to take care of the virus while the victim's body is responding to the vaccine. Preliminary results indicate that when the incubation period of the disease may be short as a result of a massive inoculation of the virus, antirabies hyperimmune serum has a definite place in the prevention of rabies. As already stated, the serum is given as soon after exposure as possible, followed by the usual vaccine injection. Hypersensitivity is a possibility—as in the case when any serum is used—and the doctor always has to consider this.

Rabies in the dog pretty much resembles the disease in man. The prodromal features of altered disposition, fever, loss of appetite, change in the tone of the bark are often so slight that only a trained observer may note them. In the typical case the prodromal phase gives way in three or four days to restlessness, agitation, tremor, and, in the "furious type," to growling and grabbing viciously at any object, animal, or person encountered. It is at this stage that the rabid dog often leaves home and travels hither and yon biting other animals and people. Death comes either during this stage of excitation or later on in the wake of paralysis. The paralytic features include immobility of the lower jaw, salivation, loss of bark, paralysis of legs, and then general paralysis, in more or less that order.

And most significant, the infectiousness of the dog or other rabid animal hinges upon the presence of the virus in the saliva. Further, in most instances the virus is present in the saliva not more than five days prior to the onset of the signs of the disease, the basic fact behind the ten-day observation period for suspected animals; that is, if rabies does not develop within this period the animal is probably not infectious.

The clinical diagnosis of rabies centers chiefly on central nervous system impairment and difficulty in swallowing following a history of exposure to a bite or lick of an animal. In the laboratory the virus can be isolated by means of neutralizing antibodies. The presence of Negri bodies in the hippocampus is conclusive proof of the disease, but their absence does not exclude a diagnosis. The standard procedure in the case of the dog is to allow the disease to become fully developed and then kill the animal, whereupon the brain is removed and examined for Negri bodies. If they are not found, a bit of mashed-up brain is injected into the brain of mice. If the virus is present these animals will die with the emergence of the telltale Negri bodies. And the fastest way to spot the virus and Negri bodies is to treat the suspected tissue with fluorescein-tagged rabies antiserum and then examining the water-rinsed specimen under the ultraviolet microscope.

A community that fails to control its canine population and fails to call for the vaccination of all dogs is providing sanctuary for the most lethal microbe known to man. These measures are musts! Also, veterinarians, laboratory workers handling cats and dogs, dog catchers, spelunkers, and everyone else engaged in any activity involving unusually high risks of exposure should be vaccinated. Finally, educating the general public to what rabies really means is essential.

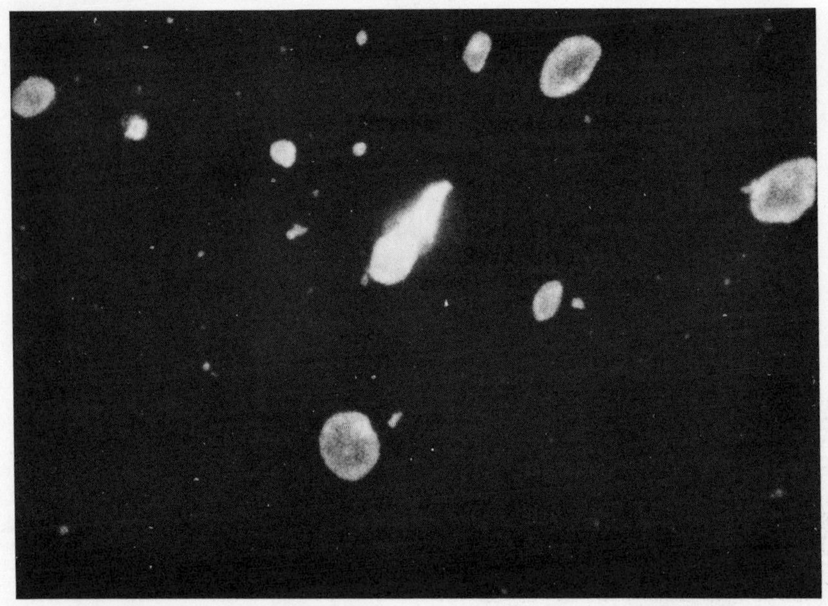

In this striking photo Negri bodies of infected mouse brain glow brightly in ultraviolet light. The tissue was prepared by adding "fluorescent antibody" (rabies antiserum "tagged" with the dye fluorescein) and then rinsing in water. The rinse removes all antibody save that which has combined with the rabies virions which compose the Negri bodies. As its name indicates fluorescein fluoresces in ultraviolet light. (Courtesy National Medical Audiovisual Center, Atlanta, Georgia.

19

The Kiss

IN 1889 THE GERMAN PHYSICIAN EMIL PFEIFFER DESCRIBED a new disease marked by fever, sore throat, swollen lymph nodes, and enlarged spleen. Quite aptly, he gave it the name "glandular fever." Twenty years later K. F. Burns called attention to the presence of strange-looking white blood cells, and in 1920 T. P. Srunt and A. S. Evans first used the term *infectious mononucleosis.* That is to say, the disease was now considered to be "infectious," and the "mononucleosis" referred to the abundance of those wayward white cells sporting a *single nucleus.* In 1923 Hal Downey described these cells in great detail and thereby set the stage for the laboratory diagnosis of the disease. To this day the so-called "Downey cell" is a hallmark of infectious mononucleosis.

The next development of significance came in 1932 when J. R. Paul and W. W. Bunnell discovered to the amazement of the medical community that the blood serum of a victim of infectious mononucleosis had a pronounced ability to clump, or agglutinate, sheep red blood cells. Not only did this prove to be an assist in the diagnosis of the disease, but also strengthened the view that infectious mononucleosis was indeed caused by a microbe—

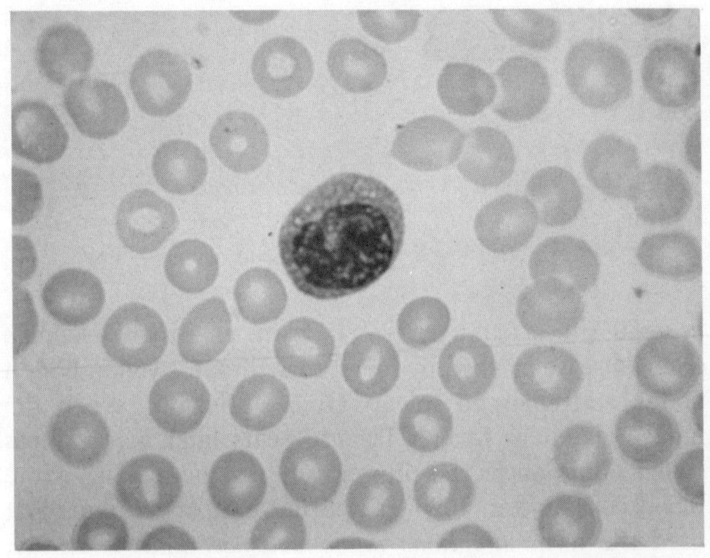

Blood picture of a patient with infectious mononucleosis. An abnormal white cell—the so-called Downey cell—stands out bright and clear in the center of the photo. (From Miale, John B.: *Laboratory Medicine-Hematology,* ed. 3 St. Louis, 1967, The C. V. Mosby Company.)

and most especially a virus since there was nothing to be seen under the microscope. Agglutination, you will recall, entails antibodies (specifically agglutinins), and these are produced in response to some microbial invader. As to why such agglutinins agglutinate sheep red cells we can only surmise that they (the red cells) chemically resemble the hypothesized virus. This agglutination test, incidentally, is referred to as the "heterophil," so-called because the agglutinins "love" (phil) "another" (hetero)—that is, the agglutinins react with antigens (in the red cells) *other* than those which brought them into existence (via plasma cell stimulation).

Following the heterophil development, Pfeiffer's glandular fever more or less passed into the limbo as far as etiology was concerned. Everything pointed to a virus, but none could be found. And then a fascinating sequence of events led Werner and Gertrude Henle, a man-and-wife research team, to what indeed may be the viral culprit.

In brief, this is the story. In 1958 an English physician, Denis Burkitt, discovered a cancerous condition among East African children that had all the earmarks of being infectious. Since the disease chiefly involved lymphoid tissue it was dubbed "Burkitt's lymphoma." Virologists throughout the world were much excited by this revelation and many set out to track down a virus. And the quest bore fruit when in 1964 M. A. Epstein and Y. M. Barr detected a herpes-like virus in tissue specimens obtained from patients with the disease. The new virus was called "EB," after its discoverers.

Whether or not the EB virus is the actual cause of Burkitt's lymphoma has yet to be demonstrated, and it was in their attempt to shed a little light on the subject that the Henles in 1968 played host to a most unexpected situation. In the course of their investigations, one of their laboratory technicians developed infectious mononucleosis—and what should her white cells disclose but the EB virus! Further, during the course of this illness antibodies against the virus appeared in her blood. And then as a sort of finale in this fortuitous chain reaction, Dr. J. C. Niederman at Yale, upon learning about this, remembered that he still had on hand twenty-nine *paired* blood samples obtained from male students who had developed infectious mononucleosis. When these were tested for antibodies against the EB virus, he found that all the *preinfection* samples were *negative* and that all the *postinfection* samples were *positive.* To date, *all* patients with infectious mononucleosis have disclosed EB antibodies, and this type of antibody persists for a much longer time than the heterophil variety mentioned earlier. This does not actually prove that the EB virus causes infectious mononucleosis, but it certainly does have critical implications. The Henles take the view that the cause is either EB or else a virus close to it.

Infectious mononucleosis is a worldwide affliction. Sporadic attacks occur chiefly between the ages of fifteen and thirty, and epidemics occur for the most part in children. Further, the college student appears to be the most likely candidate. And the feeling here has always been that prolonged "deep kissing" is the reason. In 1955, R. J. Hoagland reported that 71 out of 73 army cadets with the disease had had extensive contact six weeks or so before the onset of symptoms. Again, in 1964, Hoagland found that 7 out of 47 newlyweds developed infectious mononucleosis after about a month following their wedding.

Perhaps the most interesting survey to date in this vein was the one conducted by Kristian Ödegaard of Norway. From 1950 to 1965 he found that of 911 cases of infectious mononucleosis the average age was 17 for females and 19.5 for males . . . and of 825 patients with gonorrhea (during this period) the average female was 19 and the average male 22. One conclusion from this might be that teenagers indulge in two years of strong kissing before going on to more "fertile pastures."

The infection proper is extremely variable in its severity, as is indeed the incubation period which may run from a few days to several weeks. Typically, the initial signs and symptoms include a fever of around 100 to 103° F., malaise, sore throat and headache, all of which appear with increasing frequency during the first week or so of the illness. Once these features are fully developed the throat becomes irritated and congested and the roof of the mouth discloses circumscribed red spots in crops of ten to fifteen. In practically all cases the lymph nodes ("glands") are enlarged, and especially those of the neck and arm pit regions. In about half of the cases the spleen is tense and swollen and the liver is enlarged. The heart, kidneys and lungs may also be involved and occasionally there is a skin rash and jaundice. Also, the eyelids commonly become water-logged, resulting in a narrowed ocular aperture ("squint eyes"). *Rarely,* there may be an involuntary rapid movement of the eyeballs, poor muscular coordination, excessive sensitiveness of the skin, paralysis, or even psychotic behavior.

In most cases the cardinal features are fully developed in a week's time and persist for two to eight weeks and sometimes for several months. Very rarely death can occur as a consequence of an overwhelming bacterial infection (occasioned by lowered resistance), ruptured spleen, damaged heart, fulminating encephalitis, or suffocation, the last-named stemming from swollen vocal cords obstructing the windpipe.

The underlying mechanisms responsible for the foregoing clinical picture are apparently varied and numerous because the microscope discloses aberrations in virtually every tissue of the body. The gross changes, though, relate almost exclusively to involvement of lymphoid tissue (lymph nodes, spleen, and so on) and here we take particular note of the fact that the EB virus is *associated* not only with infectious mononucleosis but also Burkitt's lymphoma; that is, lymph tissue and the EB virus appear common to both afflictions.

The doctor makes a diagnosis of infectious mononucleosis on a "triad basis"—Downey cells, positive heterophil *and* clinical picture (namely, fever, sore throat, swollen glands, and enlarged spleen). But in an atypical case the diagnosis can be tricky, especially where the laboratory tests are not as clear cut as they could be. For instance, involvement of the abdominal lymph nodes—notably when those of the neck are not enlarged—might easily pass for appendicitis. Other diseases which confuse the situation include strep infection, trench mouth, diphtheria, typhoid fever, undulant fever, infectious hepatitis, encephalitis, polio, and leukemia.

By way of treatment, there is nothing available that can terminate an attack of infectious mononucleosis. The best advice is rest and plenty of it, particularly in the face of liver involvement. Infections due to secondary bacterial invaders call for antibiotics, and splenic rupture calls for immediate surgery. As already noted, swollen vocal cords cut off the air to the windpipe and in this eventuality a tracheotomy is lifesaving. The use of steroids (that is, cortisone and its chemical relatives) has not proved an unmixed blessing and the consensus seems to be that such medicinals should be reserved for the more severe cases.

A disease of unknown origin is more often than not a disease hard to handle and so it is here. Obviously, we all await the latest news on the EB virus, for if it is the cause and if it is indeed passed along via campus kissing, then the writing will be on the blackboard.

Bibliography

BOOKS

Beeson, P. and McDermott, W. (editors), *Cecil-Loeb; Textbook of Medicine,* 11th ed. Philadelphia: W. B. Saunders Co., 1963.

Brooks, S. M., *Basic Facts of Microbiology.* Philadelphia: W. B. Saunders Co., 1962.

———. *A Programmed Introduction to Microbiology.* St. Louis: C. V. Mosby Co., 1968.

———. Integrated Basic Science, 2nd ed. St. Louis: C. V. Mosby Co., 1966.

Burdon, K. L., and Williams, R., *Microbiology,* 6th ed. New York: Macmillan, 1968.

Burnet, F. M., *Principles of Animal Virology,* 2nd ed. New York: Academic Press, 1960.

———. and Stanley, W. M. (editors), *The Viruses.* New York: Academic Press, 1959.

Burns, J. M., *Roosevelt: The Lion and the Fox.* New York: Harcourt, Brace and World, 1956.

Burrows, W., *Textbook of Microbiology,* 19th ed. Philadelphia: W. B. Saunders Co., 1968.

Dixon, C. W., *Smallpox.* London: Churchill Ltd., 1962.

de Kruif, P., *Microbe Hunters.* New York: Harcourt, Brace and World, 1953.

Fothergill, L. D., Eastern Encephalomyelitis. *In* Beeson, and McDermott, *op. cit.*

———. Venezuelan Encephalomyelitis. *In* Beeson and McDermott, *ibid.*

Fox, J. P., St. Louis Encephalitis. *In* Beeson and McDermott, *ibid.*

———. Japanese B Encephalitis. *In* Beeson and McDermott, *ibid.*

Gebhardt, L. P., and Anderson, D., *Microbiology,* 3rd ed. St. Louis: C. V. Mosby Co., 1965.

Horsfall, F. L., and Tamm, I. (editors), *Viral and Rickettsial Infections of Man,* 4th ed. Philadelphia: J. B. Lippincott, 1965.

Hull, T. G., *Diseases Transmitted from Animals to Man,* 5th ed. Springfield, Ill.: Charles C. Thomas, 1963.

Kalter, S. S., *Procedures for Routine Laboratory Diagnosis of Virus and Rickettsial Diseases.* Minneapolis: Burgess Publishing Co., 1963.

Kerr, J. A., Yellow Fever. *In* Beeson and McDermott, *op. cit.*

Kilbourne, E. D., Influenza. *In* Beeson and McDermott, *ibid.*

———. Measles. *In* Beeson and McDermott, *ibid.*

———. Rubella. *In* Beeson and McDermott, *ibid.*

Kneeland, Jr., Y., The Common Cold. *In* Beeson and McDermott, *ibid.*

Koprowski, H., Rabies. *In* Beeson and McDermott, *ibid.*

Meiklejohn, G., Mumps. *In* Beeson and McDermott, *ibid.*

Pelczar, M. J., and Reid, R. D., *Microbiology,* 2nd ed. New York: McGraw-Hill, 1965.

Pollard, M. (editor), *Perspectives in Virology.* New York: John Wiley and Sons, Inc., 1959.

Rhodes, A. J., and Van Roogen, C. E., *Textbook of Virology.* Baltimore: William and Wilkins Co., 1962.

Robbins, S. L., *Textbook of Pathology,* 2nd ed. Philadelphia: W. B. Saunders Co., 1962.

Smadel, J. E., Smallpox. *In* Beeson and McDermott, *op. cit.*

———. Vacinnia. *In* Beeson and McDermott, *ibid.*

Smith, A. L., *Principles* of *Microbiology,* 5th ed. *St. Louis*: C. V. Mosby Co., 1965.

Smith, K. M., *Viruses.* London: Cambridge University Press, 1962.

Smith, W. (editor), *Mechanisms of Virus Infections.* New York: Academic Press, Inc. 1963.

Stanier, R. Y., Doudoroff, M., and Adelberg, E. A., *The Microbial World.* 2nd ed. Englewood Cliffs, N. J.: Prentice-Hall, 1963.

Stanley, W. M., and Valens, E. G., *Viruses and the Nature of Life.* New York: E. P. Dutton and Co., Inc., 1961.

Stents, G. S., *Molecular Biology of Viruses.* San Francisco: W. H. Freeman and Co., 1963.

Valentine, W. N., Infectious Mononucleosis. *In* Beeson and McDermott, *op. cit.*

Wagner, R. R., Herpes Simplex. *In* Beeson and McDermott, *ibid.*

Waterson, A. P., *Introduction to Animal Virology.* London and New York: Cambridge University Press, 1962.

Weidel, W., *Viruses.* Ann Arbor: The University of Michigan Press, 1960.

Weinstein, L., Poliomyelitis. *In* Beeson and McDermott, *op. cit.*

Weller, T. H., Varicella; Herpes Zoster. *In* Beeson and McDermott, *ibid.*

JOURNALS

Almeida, J. D., et al, "Electron Microscope Study of Human Warts; Sites of Virus Production and Nature of the Inclusion Bodies." *Journal of Investigational Dermatology,* 38:337, 1962.

Andrewes, C. H., "The Viruses of the Common Cold." *Scientific American,* 203:88, 1960.

———. "Tumour-viruses and virus tumors." *British Medical Journal,* 1:653, 1964.

Bell, J. A., et al, "Adenoviruses." *American Journal of Public Health,* 52:902, 1962.

Black, P. H., "Malignant Transformations in Vitro by Oncogenic Viruses." *Journal of the American Medical Association,* 206:1258, 1968.

Bloom, H. H., et al, "Relationship of Rhinovirus Infection to Mild Upper Respiratory Disease." *Journal of the American Medical Association,* 186:38, 1963.

Burkitt, D., and O'Conor, G. T., "Malignant Lymphoma in African Children—Clinical Syndrome." *Cancer,* 14:258, 1961.

Burnet, M., "The Influenza Virus." *Scientific American,* 188:27, 1953.

Cooper, L. Z., "German Measles." *Scientific American,* 214:30, 1966.

Dalldorf, G., "The Coxsackie Viruses." *Bulletin of the New York Academy of Medicine,* 26:329, 1950.

Dean, D. J., "Pathogenesis and Prophylaxis of Rabies in Man." *New York Journal of Medicine,* 63:3507, 1963.

Deinhardt, F., "Mumps." *Medical Science,* 11:827, 1962.

Douglas, R. G., et al, "Exposure to Cold Environment and Rhinovirus Common Cold." *New England Journal of Medicine,* 279:742, 1968.

Enders, J. F. et al, "Studies on an Attenuated Measles-virus Vaccine." *New England Journal of Medicine,* 263:153, 1960.

Franhel, J. W. et al, "Measles Vaccination." *Proceedings of the Society for Experimental Biology and Medicine,* 110:154, 1962.

Gebhardt, L. P., "Overwintering of Western Equine Encephalitis Virus." *Proceedings of the Society for Experimental Biology and Medicine,* 104:695, 1960.

Henle, W., et al, "Antibody Coating and Agglutination of Virus Particles Separated from the EB Live of Burkitt Lymphoma Cells." *Journal of Bacteriology,* 92:269, 1966.

Hilleman, M. R., "Respiratory Viruses and Respiratory Virus Vaccines." *American Review of Respiratory Diseases,* 87:165, 1963.

Hoagland, R. J., "The Transmission of IM." *American Journal of Medical Science,* 229:262, 1955.

Holland, J. J., "Enterovirus Entrance into Specific Host Cells and Subsequent Alterations of Cell Protein and Nucleic Acid Synthesis." *Bacteriological Reviews,* 28:3, 1964.

Horne, R. W., "The Structure of Viruses." *Scientific American,* 208:48, 1963.

Horsfall, F. L., "Viral Infections of the Respiratory Tract." *American Review of Respiratory Diseases,* 80:315, 1959.

Hortsman, D. M., et al, "Immunization of Preschool Children with Oral Poliovirus Vaccine (Sabin)." *Journal of the American Medical Association,* 178:693, 1961.

Jacob, F., and Wollman, E., "Viruses and Genes." *Scientific American,* 204:92, 1961.

Jones, J., "The Sexual Life of a Mosquito." *Scientific American,* 218:108, 1968.

Kaufman, H. E., "Treatment of Ocular Herpes Simplex: Use of Corticosteroids and IDU." *Southern Medical Journal,* 57:163, 1964.

Kornberg, A., "The Synthesis of DNA." *Scientific American,* 218:64, 1968.

La Boccetta, A. C., et al, "Measles Encephalitis." *American Journal of the Diseases of Childhood,* 107:247, 1964.

Langmuir, A. D., "Asian Influenza in the United States." *Annals of Internal Medicine,* 49:483, 1958.

Lepow, M. L., et al, "Immunization Levels After Oral Poliovaccine." *Journal of the American Medical Association,* 187:749, 1964.

Maramorosch, K., "Friendly Viruses." *Scientific American,* 203:138, 1960.

Melnick, J. L., "Enteroviruses." *Scientific American,* Vol. 200, 1959.

Merigan, T. G., et al, "Modified Skin Lesions During Concurrent Varicella and Measles Infections." *Journal of the American Medical Association,* 204:123, 1968.

Miller, A. H., "Relationship of Surgery of the Nose and Throat to Poliomyelitis." *Journal of the American Medical Association,* 150:532, 1952.

Miller, G. D. and Tindall, J. P., "Hand-Foot-and-Mouth Disease." *Journal of the American Medical Association,* 203:107, 1968.

Mims, C. A., "Aspects of the Pathogenesis of Virus Diseases." *Bacteriological Reviews,* 28:30, 1964.

Molomut, N. and Padnos, M., "Cytopathic Effects of M-P Virus on Lymphoid Cancer in Vivo." *Journal of the American Medical Association,* 204:218, 1968.

Morgan, C., et al, "Structure and Development of Viruses Observed in the Electron Microscope. (Influenza Virus.)" *Journal of Experimental Medicine,* 104:171, 1956.

———. "Structure and Development of Viruses Observed in the Electron Microscope. (Vaccinia and Fowlpox Viruses.)" *Journal of Experimental Medicine,* 100:301, 1954.

———. "Structure and Development of Viruses Observed in the Electron Microscope. (RI-APC Group.)." *Journal of Biophysical and Biochemical Cytology,* 2:351, 1956.

———. "Intracellular Crystals Associated with Viral Development." *Transactions of the Academy of American Physicians,* 71: 281, 1958.

Niederman, J. C., et al, "Infectious Mononucleosis." *Journal of the American Medical Association,* 203:205, 1968.

Pandit, C. G., "Newly Recognized Viral Diseases—With Special Emphasis on Hemorrhagic Types Found in Asia." *American Journal of Public Health,* 50:46, 1960.

Rawls, W. E., et al, "Herpes Type 2: Association with Carcinoma of the Cervix." *Science,* 161: 1255–1256, 1968.

Sabin, A. B., "Paralytic Consequences of Poliomyelitis Infection in Different Parts of the World and in Different Population Groups." *American Journal of Public Health,* 41:1215, 1951.

Schaeffer, M., et al, "American Arthropod-borne Virus Problems." *American Journal of Public Health,* 48:336, 1958.

Southam, C. M., "Viruses in the Field of Cancer." *Modern Medicine,* 28:74, 1960.

Stamm, D. D., "Studies on the Ecology of Equine Encephalomyelitis." *American Journal of Public Health,* 48:328, 1958.

Weibel, R. E., et al, "Durable Immunity Two Years after Administration of Ender's Live Measles Vaccine with Immune Globulin." *New England Journal of Medicine,* 270: 172, 1964.

Wilson, J. L., "Relationship of Tonsillectomy to Incidence of Poliomyelitis." *Journal of the American Medical Association,* 150:539, 1952.

Wood, W. B. and Edgar, R. S., "Building a Bacterial Virus." *Scientific American,* 217:61, 1967.

Wyckoff, R. W. G., "Electron Microscopic Study of Viruses." *Journal of the American Medical Association,* 136:1081, 1948.

Yoeli, M., "Animal Infections and Human Disease." *Scientific American,* 202:161, 1960.

Index